THE CRUDE OIL EXPORT BAN: HELPFUL OR HURTFUL?

HEARING

BEFORE THE

SUBCOMMITTEE ON TERRORISM, NONPROLIFERATION, AND TRADE

OF THE

COMMITTEE ON FOREIGN AFFAIRS
HOUSE OF REPRESENTATIVES

ONE HUNDRED FOURTEENTH CONGRESS

FIRST SESSION

APRIL 14, 2015

Serial No. 114–22

Printed for the use of the Committee on Foreign Affairs

Available via the World Wide Web: http://www.foreignaffairs.house.gov/ or
http://www.gpo.gov/fdsys/

U.S. GOVERNMENT PUBLISHING OFFICE

94–178PDF   WASHINGTON : 2015

COMMITTEE ON FOREIGN AFFAIRS

EDWARD R. ROYCE, California, *Chairman*

CHRISTOPHER H. SMITH, New Jersey
ILEANA ROS-LEHTINEN, Florida
DANA ROHRABACHER, California
STEVE CHABOT, Ohio
JOE WILSON, South Carolina
MICHAEL T. McCAUL, Texas
TED POE, Texas
MATT SALMON, Arizona
DARRELL E. ISSA, California
TOM MARINO, Pennsylvania
JEFF DUNCAN, South Carolina
MO BROOKS, Alabama
PAUL COOK, California
RANDY K. WEBER SR., Texas
SCOTT PERRY, Pennsylvania
RON DeSANTIS, Florida
MARK MEADOWS, North Carolina
TED S. YOHO, Florida
CURT CLAWSON, Florida
SCOTT DesJARLAIS, Tennessee
REID J. RIBBLE, Wisconsin
DAVID A. TROTT, Michigan
LEE M. ZELDIN, New York
TOM EMMER, Minnesota

ELIOT L. ENGEL, New York
BRAD SHERMAN, California
GREGORY W. MEEKS, New York
ALBIO SIRES, New Jersey
GERALD E. CONNOLLY, Virginia
THEODORE E. DEUTCH, Florida
BRIAN HIGGINS, New York
KAREN BASS, California
WILLIAM KEATING, Massachusetts
DAVID CICILLINE, Rhode Island
ALAN GRAYSON, Florida
AMI BERA, California
ALAN S. LOWENTHAL, California
GRACE MENG, New York
LOIS FRANKEL, Florida
TULSI GABBARD, Hawaii
JOAQUIN CASTRO, Texas
ROBIN L. KELLY, Illinois
BRENDAN F. BOYLE, Pennsylvania

AMY PORTER, *Chief of Staff*   THOMAS SHEEHY, *Staff Director*
JASON STEINBAUM, *Democratic Staff Director*

———

SUBCOMMITTEE ON TERRORISM, NONPROLIFERATION, AND TRADE

TED POE, Texas, *Chairman*

JOE WILSON, South Carolina
DARRELL E. ISSA, California
PAUL COOK, California
SCOTT PERRY, Pennsylvania
REID J. RIBBLE, Wisconsin
LEE M. ZELDIN, New York

WILLIAM KEATING, Massachusetts
BRAD SHERMAN, California
BRIAN HIGGINS, New York
JOAQUIN CASTRO, Texas
ROBIN L. KELLY, Illinois

CONTENTS

THE CRUDE OIL EXPORT BAN: HELPFUL OR HURTFUL?

TUESDAY, APRIL 14, 2015

HOUSE OF REPRESENTATIVES,
SUBCOMMITTEE ON TERRORISM, NONPROLIFERATION, AND TRADE,
COMMITTEE ON FOREIGN AFFAIRS,
Washington, DC.

The committee met, pursuant to notice, at 10:15 a.m., in room 2172 Rayburn House Office Building, Hon. Ted Poe (chairman of the subcommittee) presiding.

Mr. POE. The subcommittee will come to order. Without objection, all members may have 5 days to submit statements, questions, extraneous materials for the record subject to the length limitation in the rules.

The United States is now the largest crude oil producer in the world. We have more oil than we can refine or store. The majority of U.S. refineries were built to handle heavy, sour crude, but oil production is light, sweet crude. The United States' refineries cannot keep up with the new production.

Normally producers would simply pump oil into storage containers, but experts say those storage tanks could fill up before the end of this very month. Instead of exporting excess oil like producers get to do in other nations, the ban is already forcing U.S. oil producers to leave oil in the ground and lay off workers. About 50 percent of the working rigs in my home state of Texas have had to shut down in just the last 6 months. Seventy thousand oil workers have been laid off since Thanksgiving.

The solution to this problem is clear: Export crude oil; have the ban lifted so that it can be exported. Critics of lifting the ban are afraid the United States' oil exports will lead to higher domestic gas prices. However, many studies have debunked this myth. Gas prices are more closely linked to the international market, or Brent Price, than the domestic price of crude because refined products like gasoline are traded freely on the international market. So the more crude oil we have, the more we can put on the international market, and the lower the international price of crude. The lower the international price of crude the lower the price of gas for America.

A Rice University study released in March 2015 reviewed previous studies that examined the impact of removing the ban on gas prices. They found that all studies underscore that lifting of the export ban will not translate into higher gasoline prices. In fact, stud-

(1)

ies generally project gasoline prices in the U.S. will fall once the ban is lifted.

U.S. crude entering the global market will increase the international oil supply and decrease the price of gas. The only thing the studies do not agree on is just how much the gas prices will drop. Lifting the ban will also lead to more jobs and higher GDP. An IHS study predicts crude oil exports would support nearly 300,000 jobs by 2018. Removing the export ban would add 26 billion to the GDP per year and improve labor income about $158 per year on average.

As it improves the U.S. economy, removing the ban will also improve our national security. The original purpose of the ban put back in 1973 was to insulate the United States from the volatility of the international oil market. Ironically, today the ban exposes the United States' market to volatility. If ISIS continues to wreak havoc and disrupt oil prices in places like Libya and Iraq, having more U.S. crude oil on the market would help prevent a spike in the price of crude oil and gas prices. Lifting the ban would free us up to help our allies.

Europe gets 40 percent of its oil from Russia. Exporting crude oil would give the Europeans an alternative to having to depend on Russia. It would also increase our influence in Asia. Japan and South Korea partly rely on crude oil from Iran to satisfy the growing energy consumption. U.S. exports can help diminish that reliance.

It is ironic to me, with the so-called deal with the Iranians, that it is now the U.S. Government's long-term policy to allow Iran to export crude oil and inject billions of dollars in their own economy. At the same time, it is still the U.S. Government's policy to prohibit American producers from doing the same. It seems to me what is good for the Iranian oil exports, should be the same deal that the United States' oil producers get.

U.S. exports offer a stable energy to our allies and decrease their reliance on dictators and state sponsors of terror. Lifting the ban shows the U.S. is serious about supporting free markets around the world. We criticize China for not exporting rare earth materials and yet we are not exporting crude oil. Removing the ban will give us more credibility when we criticize export bans in other nations.

All in all, it is time we remove the crude oil export ban. Exporting crude oil will lower gas prices, increase American jobs and strengthen our national security. And that is just the way it is, to coin a phrase.

I will now yield to the ranking member, Mr. Keating from Massachusetts for his opening statement.

Mr. KEATING. Well, thank you, Chairman Poe. And I would like to thank our witnesses, my colleagues, for being here. I feel a little bit relieved, because I am juggling between an important bill in Homeland Security today. And it is great to see the chairman here, and I think it also shows how important he believes this issue is for his district.

And I look forward to an informative discussion today concerning what our witnesses see as the costs and benefits of lifting the current ban on exporting U.S. crude oil. It is vital that we consider the economic, environmental, and foreign policy implications of our dependence on fossil fuels and of the imports by the United States and our allies of oil and gas from volatile regions such as the Middle East and Eurasia.

And while some oil companies advocate for diversification of energy sources on geopolitical grounds, I have seen instances of some oil companies actually obstructing renewable energy technologies, thereby undercutting their own argument, some of them, for diversification. I am concerned about the environmental consequences of ending the crude oil export ban and look forward to hearing from our witnesses in that regard.

Lifting the export ban would cause the domestic price of crude oil to increase, many say, which would then lead to an increase in the production of U.S. crude oil. An increase in the production of domestic crude oil would have serious negative environmental impacts as well.

For example, rising domestic crude oil production would heighten the risk of spills in transporting crude oil by pipeline, rail, truck, barge or tanker, and the negative health and environmental impacts of those spills are a concern as well. In addition to expanded domestic crude oil production, it would likely cause a significant increase in the release of carbon dioxide which contributes to climate change.

The environmental cost of producing crude oil and continuing to rely on fossil fuels underscores that U.S. energy policy must seek to diversify our sources of energy and increase the production of wind, solar and other forms of cleaner, renewable energy. So I hope that is part of the discussion today as well and with that I yield back, Mr. Chair.

Mr. POE. I thank the ranking member. All members may file their statements. Without objection, all the witnesses' prepared statements will be made a part of the record, and I will now introduce our first two witnesses on the first panel, both Members of Congress.

Congressman Michael McCaul represents the 10th congressional district in the great state of Texas. Congressman McMaul, as already stated, is chairman of the House Committee on Homeland Security.

We also have Congressman Joe Barton. He is from Texas as well and represents the 6th district. Congressman Barton is the chairman emeritus of the House Energy and Commerce Committee. Both representatives have introduced legislation to lift the crude oil export ban that is pending before this subcommittee.

I appreciate both of you being here. I will say for the record, Mr. McCaul does have a markup very soon. And without objection as soon as you testify you may leave, and Mr. Barton will stick around and answer all of the questions that would have been addressed to you after he testifies.

Mr. McCaul?

STATEMENT OF THE HONORABLE MICHAEL MCCAUL, UNITED STATES HOUSE OF REPRESENTATIVES

Mr. McCAUL. Thank you, Mr. Chairman, and I am sure Mr. Barton will do a fine job answering questions directed at me. I must say we mentioned how intimidating it is to be sitting in the well here with the chair and ranking member so high up on the dias, and I can only imagine what the criminal defendants in your courtroom, how they must have felt before they went before you. Fortunately, I never had that opportunity.

So I just need to say thank you for having this hearing and markup on this very important issue. It is long overdue to lift the 40-year-old ban on crude oil exports. And I think it is fitting that we should have this discussion in the Foreign Affairs Committee because crude oil exports is a major foreign policy issue.

Around the world our friends and allies are looking for a stable and reliable supply of American energy. And countries like Russia abuse their status as a dominant energy supplier to bully their neighbors in Europe and Central Asia, while supply disruptions from places like Iran and Libya leave the global oil markets vulnerable to price spikes. The geopolitical benefits of American energy exports as a diplomatic tool will both make us stronger economically and provide critical support to our partners around the world.

But don't take it from me, take it from the European Union's trade negotiators who leaked a memo last year acknowledging that the crisis, and this is their quote: ''The crisis in Ukraine confirms the delicate situation faced by the EU with regard to energy independence.''

And then they urged the administration privately to lift, they said, ''Lifting bilateral restrictions on gas and crude oil to increase security and instability through open markets.'' To lift these restrictions to increase security and stability—this is the European Union. This is not you or I talking on the Republican side of the aisle.

Or take it from Larry Summers, President Obama's former director of the National Economic Council and President Clinton's former Treasury secretary, hardly a Republican from Texas—I put that one in my script. He argues that if we wish to have more power and influence in the world in support of our security interest and in support of our values, and if we wish to have an influence, that we pay for with neither blood nor taxes, I do not see a more constructive approach than permitting the export of fossil fuels. Larry Summers. Of course the ranking member knows him well having represented the Boston area, and of course Harvard University being there. Some of my colleagues—I yield

[The prepared statement of Mr. McCaul follows:]

Prepared Testimony for Rep. McCaul
House Foreign Affairs Committee
Subcommittee on Terrorism, Nonproliferation, and Trade
"The Crude Oil Export Ban: Helpful or Hurtful?"
14 April 2015

Chairman Poe, Ranking Member Keating and Members of the Subcommittee, thank you for the opportunity to discuss a priority for our country: lifting the forty year old ban on crude oil exports.

It's fitting that we should have this discussion in the Foreign Affairs Committee because crude oil exports is a major foreign policy issue.

Around the world, our friends and allies are looking for a stable and reliable supply of American energy. Countries like Russia abuse their status as a dominant energy supplier to bully their neighbors in Europe and Central Asia, while supply disruptions from places like Iran and Libya leave the global oil market vulnerable to price spikes. The geopolitical benefits of American energy exports as a diplomatic tool will both make us stronger economically and provide critical support to our partners around the world.

But don't take it from me. Take it from the European Union's trade negotiators who leaked a memo last year acknowledging that the "crisis in Ukraine confirms the delicate situation faced by the EU with regard to energy dependence" and urged U.S. support for "lifting bilateral restrictions on gas and crude oil... to increase security and stability through open markets."[1]

Or take it from Larry Summers, President Obama's former Director of the National Economic Council and President Clinton's former Treasury Secretary – hardly a Republican from Texas – who argues that "If we wish to have more power and influence in the world, in support of our security interests, and in support of our values... and if we wish to have an influence that we pay for with neither blood nor taxes, I do not see a more constructive approach than permitting the export of fossil fuels."[2]

Some of my colleagues who are skeptical about lifting the ban contend that allowing crude exports could increase gas prices. They also argue that keeping our crude oil here at home makes us more energy secure. Allow me to address both of these points.

---

[1] Council of the European Union. Non-paper on a Chapter on Energy and Raw Materials in TTIP, May 27, 2014.
[2] Brookings Institution, "Larry Summers Argues Case for Lifting the Crude Oil Export Ban," September 9, 2014.

The nonpartisan Government Accountability Office found that removing crude oil export restrictions will actually decrease gasoline prices by 1.5 to 13 cents per gallon for American consumers.[3] The Energy Information Administration explained this somewhat counterintuitive phenomenon by pointing out that "the effect that a relaxation of current limitations on U.S. crude oil exports would have on U.S. gasoline prices would likely depend on its effect on international crude oil prices... rather than its effect on domestic crude prices."[4] We already allow for the free trade of gasoline – there is no reason why crude should be treated differently.

Keeping the ban in place will in fact make us less energy secure and restrict economic growth. Without the option to export to foreign markets, our producers will continue to be forced to sell their crude oil at an artificially discounted rate, which is already causing them to cut back production. This is a real problem in states like mine, where small independent producers are laying off workers. But it also holds back growth in states that produce little to no energy at all because of the effect on other industries that support crude oil producers.

Moreover, U.S. refiners are not fully optimized to process the explosive growth in domestic production of light sweet crude in states like Texas and North Dakota. Rather, they are configured to refine heavy crude from countries such as Canada and Mexico. Allowing for the free trade of crude oil will make the market more efficient by correcting this producer-refiner mismatch.

In conclusion, Mr. Chairman, the crude oil export restrictions enacted in the wake of the 1970s Arab Oil Embargo are no longer justified given today's market conditions. I urge this subcommittee to repeal the export ban in its entirety, but I think it's critical that Congress create a safety valve that ensures the President has the ability to restrict exports in the case of an unforeseen national emergency. My bill, H.R. 156, the *Crude Oil Export Act*, which has been referred to this subcommittee, has such a provision.

Thank you again for the opportunity to discuss my legislation. I look forward to working in a bipartisan fashion to address this critical issue.

---

[3] Government Accountability Office, Report to the Ranking Member, Committee on Energy and Natural Resources, U.S. Senate, "Changing Crude Oil Markets", September 2014, page 16.
[4] U.S. Energy Information Administration, "What Drives U.S. Gasoline Prices?" October 2014, page 3.

Mr. KEATING. Mr. Chairman, you have gone there too, as a Texan.

Mr. MCCAUL. Some of my colleagues are skeptical. I don't question their intent, but they are skeptical about lifting the ban. They contend that allowing crude oil exports could increase gas prices. They also argue that keeping our crude oil here at home makes us more energy secure.

Allow me to address both of these points. First, the nonpartisan Government Accountability Office found that removing crude oil export restrictions will actually decrease gasoline prices by 1.5 to 13 cents per gallon for American consumers. The Energy Information Administration explained this somewhat counterintuitive phenomenon, pointing out that the effect that a relaxation of current limitations on U.S. crude oil exports would have on U.S. gasoline prices would likely depend on its effect on international crude oil prices rather than its effect on domestic crude oil prices.

We already allow for the free trade of gasoline so there is no reason why crude oil should be treated differently. Keeping the ban in place will in fact make us less energy secure, restrict economic growth, and without the option to export to foreign markets our producers will continue to be forced to sell their crude oil at an artificially discounted rate—which is already causing them, as the chairman mentioned, to cut back production by 50 percent. This is a real problem in states like mine where independent small producers are laying off workers. It also holds back growth in states that produce little to no energy at all because of the effect on other industries that support crude oil producers.

Moreover, U.S. refineries are not fully optimized to process the explosive growth in domestic production of light, sweet crude in states like Texas and North Dakota, rather, they are configured to refine heavy crude from countries such as Canada and Mexico. Allowing for the free trade of crude oil will make the market more efficient by correcting this producer/refiner mismatch.

In conclusion, Mr. Chairman, the crude oil export restrictions enacted in the wake of the 1970s era oil embargo are no longer justified given today's market conditions. And as the chairman of Homeland Security, if this has a devastating or detrimental impact on ISIS and Iran, I think that would be a positive thing in our foreign policy and homeland security and I urge the subcommittee to repeal the export ban in its entirety as my bill does. But I think it is critical that Congress create a safety valve that ensures the President has the ability to restrict our exports in the case of unforeseen national emergencies. My bill, H.R. 156, the Crude Oil Export Act, which has been referred to this subcommittee, has such a provision.

So I want to thank you again for drawing attention to the issue. It is an important issue. I think there will be a healthy debate on the committee that I serve on as well on this issue and I look forward to the day that they are both marked up and sent to the House floor for a vote. And with that I yield back.

Mr. POE. I thank the chairman, Chairman McCaul, and you can leave whenever you need to get to your committee. However, hopefully Mr. Keating the ranking member will stick around here for awhile.

The chair now recognizes the gentleman from Texas.

Mr. MCCAUL. When we have votes I will need the ranking member on the markup. So with that I yield back.

Mr. POE. All right.

STATEMENT OF THE HONORABLE JOE BARTON, UNITED STATES HOUSE OF REPRESENTATIVES

Mr. BARTON. Well, thank you, Mr. Chairman, and thank you Ranking Member Keating, Mr. Sherman, Mr. Cook, Mr. Perry, Mr. Ribble, and Mr. Rohrabacher for attending this hearing. A special thanks to you, Chairman, for scheduling it. I am glad that your committee doesn't have a limitation on the number of Texas witnesses. Most committees do, but I am glad that you are ecumenical. I am going to submit my written statement for the record and just speak extemporaneously.

My bill is a very simple bill. It is a page long. It repeals the ban on crude oil exports that was first established back in 1975. It repeals it in its entirety. It is not phased in and phased out. It just repeals it very simple, and then it requires a study of what to do, if anything, with the strategic petroleum reserve which we established at about that same time as a buffer against any future Arab oil embargoes.

Back in 1975 when the ban was put in place the world was a different place. U.S. production was declining and imports were increasing. We were in a bad economic situation. And it was felt at that time that oil was of such strategic importance that it should not be allowed to be exported. It is the only energy commodity that is so restricted. We export coal. We export natural gas. We even export wood chips and electricity. But we don't export crude oil.

Now oil is fungible, Mr. Chairman and members of this subcommittee. There are differences in viscosity and sulfur content, but basically oil is oil. It can go anywhere. If we had a barrel of West Texas intermediate and we had a barrel of Saudi light, an expert with some testing could tell the difference but nobody on this committee could tell the difference.

So the reason that I think we need to repeal the ban is pretty straightforward. U.S. oil production is increasing. It is over 10 million barrels a day and going up at least for the time being. We produce more oil than Saudi Arabia which is number two, or Russia that is number three. If we were to repeal the ban on crude oil exports, we would allow U.S. producers to sell their oil to any willing buyer whether it was domestic or foreign.

What difference does that make you might ask? Well, it is pretty straightforward. Because we have a ban on producers selling on the world market today they can only sell to domestic refiners. Now that is automatically a bad thing. I am a friend of the U.S. refining industry. But because they can't sell on the world market and because there is such a glut of oil being produced in the United States, U.S. refiners don't have to pay the world market price. So they get a discount, what I call a ''domestic discount.'' And again, that in and of itself is not automatically a bad thing.

But the refiners take this discount, they offer our producers less, they refine it and then they export on the world market. We are exporting about 3 million barrels a day of refined products. Those products are sold at world market prices, but the producer who is producing the crude oil is not getting the world price.

Now that discount has varied over time. Right now it is about $10 a barrel. It has been as high as 30. If we repeal the ban, Mr. Chairman, on crude oil exports that discount disappears. Now that is not necessarily a bad thing for U.S. refiners, but it will be a good thing for U.S. consumers because putting more U.S. oil or any U.S. oil on the world market will tend to depress or at least stabilize world market prices, and that will result in lower pump prices over time for our consumers whether they be in Massachusetts, California, Texas or wherever.

So Mr. Chairman, my time is about to expire but let me simply say, this is a win-win. It is a win for the consumer. It is a win for the producer. It is a win for the strategic interests of the United States, and it puts pressure on the Saudis and the Russians which are not, at least in the case of Russia is not always our friend.

With that, my time is expired. I would be happy to answer any questions. But thank you for the hearing and thank all the members for their attendance.

[The prepared statement of Mr. Barton follows:]

Chairman Poe, Ranking Member Keating, Members of the Committee - Thank you for inviting me here today to testify on my bill HR 702, a straightforward piece of legislation that repeals authority to restrict the export of crude oil. I would like to thank those of you who are already cosponsors and look forward to winning the support of others in the future.

My colleague Chairman McCaul has introduced similar legislation with the same objective—to remove an outdated law that stands contrary to America's free market ideals and constrains our nation's potential economic growth. It is my understanding that Chairman McCaul will discuss the geopolitical benefits of removing the ban, of which there are many. But first, a little background:

The Energy Policy and Conservation Act of 1975 was crafted at a time when OPEC delivered 65% of US oil imports and the idea of an energy secure America was unimaginable.[1] Three assumptions underpinned that legislation: domestic production was dwindling; consumption was rising; and reliance on foreign sources of oil was increasing. In response to the Arab oil embargo Congress enacted a suite of policies to reduce US reliance on foreign imports including restrictions on exporting all domestically produced oil and petroleum products, plus a complicated system of price controls. We repealed price controls 34 years ago, lifted restrictions on product exports and have subsequently become a net exporter of these products, but the ban on crude oil exports still stands. We are the only member of the Organization for Economic Cooperation and Development (OECD) and the International Energy Agency (IEA) with an outright ban on exporting domestically produced crude oil.

In December, the US surpassed Saudi Arabia as the world's largest producer of petroleum products. 96% of our growth in production since 2011 has come from a certain type of oil known as light sweet crude.[2] Our refining system, one of the most technologically advanced and efficient in the world, is configured to refine primarily heavy crude. The glut of light oil we see stocking up in storage in Cushing, Oklahoma and elsewhere across the nation has nowhere to go. Now, with the price of oil nearly half what it was this time last year, Texas producers are especially feeling pinched. The number of oil rigs in Texas actively drilling has dropped more than 50% since October when we had 1,609 active rigs.[3] Today we have 760, the lowest weekly total since December 2010. That's a lot of production going offline, costing thousands of jobs and untold economic slowdown. Due to declining production, Texas lost 40% of its tax revenue from oil and gas in January 2015.[4] This ban hurts our constituents, our cities, and our country.

---

[1] Testimony of Adam Sieminski before the House Energy and Commerce Committee, December 11, 2014.
[2] See "Increases in U.S. crude oil production come from light, sweet crude from tight formations," U.S. Energy Information Administration
[3] See "North America Rig Count," Baker Hughes, April 10, 2015.
[4] See "Oil Price decline leads to lower tax revenue in top oil-producing states," U.S. Energy Information Administration.

Removing the crude export ban will be good for the country, increasing domestic production, raising GDP, and increasing federal revenues approximately $35 billion higher in 2020 than without shale development.[5] Without the ban, nearly 630,000 jobs will be added at peak production in 2019 and household incomes will see a real increase of $2,000 to $3,000 per household in 2025.[6]

Studies show that removing the ban would decrease the price of gasoline we pay at the pump up to 13 cents per gallon, that's real money to real families.[7] EIA has shown that the price we pay at the pump is based on the international price of oil and it follows that more supply in the international market will push the price downward for folks at home.

There is broad agreement that maintaining the ban is bad for business and bad for America. It is up to this Congress to examine the issue and move towards a better policy that reflects the reality of America today, not the America of 1975. It is a win, win, win—internationally, domestically, and economically.

---

[5] See "The Economic and Budgetary Effects of Producing Oil and Natural Gas from Shale," Congressional Budget Office, December 9, 2014.
[6] See "Lifting the Crude Oil Export Ban: The Impact on U.S. Manufacturing," The Aspen Institute, October 2014.
[7] See "U.S. Crude Oil Export Policy: Background and Considerations," Congressional Research Service, December 2014.

Mr. POE. I thank the gentleman for his testimony. Are there any questions from members of the panel?

Mr. ROHRABACHER. Mr. Chairman. Mr. Rohrabacher. And let me just congratulate our colleagues, Mr. McCaul and Barton, moving forward like this. This is a really important issue and in the long run it is going to have a very positive impact on our people. And it is time that government got off our hind end and just got out there and got some things done and this is one of the things we could do to make things better. So thank you very much for your leadership.

Mr. POE. Thank you. The gentleman yields back.

Anyone else?

I want to thank you, Mr. Barton, for being here. You are excused. You do not have to stay.

Mr. BARTON. Thank you, sir. And let me simply say I am willing to discuss this one-on-one with any of the members of your sub-committee or the full committee. I do sincerely appreciate you having a hearing and I look forward to discussing this in the future. Thank you. Thank you.

Mr. POE. Thank you, Mr. Barton.

We will get ready for our next panel, if they will come up. I want to thank our panelists for being here. I will introduce each one of you, and then we will go in the same order for your testimony; and limit your testimony to 5 minutes please.

Mr. Jason Grumet is the founder and president of the Bipartisan Policy Center. Previously Mr. Grumet founded and directed the National Commission on Energy Policy.

Ms. Elizabeth Rosenberg is director of the Energy, Economics and Security Program at the Center for a New American Security. Prior to this position, Ms. Rosenberg served as a senior advisor at the U.S. Department of the Treasury to the assistant secretary for Terrorist Financing and Financial Crimes, and then to the under-secretary for Terrorism and Financial Intelligence.

Mr. Jason Bordoff is professor of Professional Practice in International and Public Affairs and founding director of the Center on Global Energy Policy at Columbia University. Before joining the Columbia faculty, Mr. Bordoff served as special assistant to the President and senior director for energy and climate change on the staff of the National Security Council.

Dr. Stephen Kretzmann is the founder and executive director of Oil Change International. Mr. Kretzmann has worked on environmental and social issues around the global fossil fuel industry for the last 25 years.

Mr. Grumet, we will start with you, and you have 5 minutes.

STATEMENT OF MR. JASON GRUMET, FOUNDER AND PRESIDENT, BIPARTISAN POLICY CENTER

Mr. GRUMET. Thank you very much, Chairman Poe, Ranking Member Keating, Mr. Sherman. I will thank you again, Chairman Poe, Ranking Member Keating, Mr. Sherman, Mr. Ribble, Mr. Rohrabacher, the committee, for the privilege to be with you this morning. As I hope my testimony reveals, the Bipartisan Policy Center supports efforts to lift restrictions on crude oil exports.

In the broadest sense, this ban is a 40-year-old anachronism. It was passed at a moment of significant national weakness. The irony is that this policy is now inhibiting one of our nation's greatest strengths. Our energy abundance has profound potential to continue to accelerate our economic recovery, to strengthen our interests internationally, and we do believe it is time for it to be reconsidered and lifted. Left unaddressed, the policy will undermine domestic production and it will weaken our recovery.

But more relevant, I think, to this committee, keeping U.S. resources and market power on the sidelines empowers our adversaries to use their energy as a weapon. It diminishes our ability to produce a myriad, and pursue a myriad, of policy imperatives, and it undermines our ability and credibility to advocate for free trade in open markets.

What I would like to do is try to summarize and frame a few ideas around the economy and then around foreign policy, and if the clock allows a few words about maybe a path forward. And let me just begin in the crucible of at least the political conversation which of course is gas prices.

Inevitably, the political debate will ultimately come down to assertions and perceptions about the impact of any policy change on the price at the pump. But fundamentally, consumers are really somewhat on the sidelines in this debate. This debate is fundamentally a commercial dispute between producers who want access to the prices in a global market and refiners who are enjoying, as the Congressman said, the discount of a lower cost crude supply.

For many, I think, the expectations about consumer impact rest on the misconception that refiners through some imagination of altruism are going to pass on these savings to consumers and drivers. But this simply is not how competitive markets work. Refiners appropriately seek the highest price for their product capturing any windfall for their shareholders. Because gasoline and refined products are of course sold globally, it is the global price that affects us here in the U.S. and not the price of domestic crude.

And there have been a number of studies which I hope we will talk about a little bit that basically endorse this proposition. The group, IHS, did a detailed assessment which asserted that prices in the U.S. would fall by 8 to 12 cents a gallon. Mr. Bordoff has done fine work that I believe suggests that the price could go down by up to 8 cents a gallon. Rice University, the Energy Information Administration, and the Resources for the Future, have all essentially confirmed the same idea: That lifting the mandate will increase global production, and in doing so, add supply to the market which will create reductions in price and more resiliency.

And while none of us can pretend to know exactly what the extent of those benefits are—and I would suggest, Mr. Poe, that if we did I would ask you to pause the hearing so we could all run and call our brokers—it is pretty clear that adding supply to the global market is going to have a beneficial impact on prices.

One last point about economics and that is the simple but obvious point about jobs. The abundance in energy has been a dramatic, I think, improvement to recent economic struggles and by increasing production we will in fact increase the availability of good high paying jobs in this country. It is true that the market for jobs around energy production has diminished as prices have gone down, but consumers have had that benefit of lower prices. The double whammy of the ban is that it depresses economic productivity at home without in fact providing those consumer benefits, and that I think is the reason economically why we believe it is a barrier to progress.

Let me turn now to trade and then a moment on how we project power. I think the U.S. has righteously decried resource nationalism for decades and protectionism that has inhibited and hindered global energy markets. And until recently our four decade old ban was essentially, I would like to think, was kind of a quaint hypocrisy. It was an aberration in policy, but it really had no impact on markets because we had no excess capacity in fact to share with the world. This has now of course changed, and for the Congress to perpetuate the ban at this moment I think would in fact undermine our credibility in promoting open markets.

Finally, talking about the impact on foreign policy, the ban simply empowers our adversaries. Absent spare capacity in the global market, any unanticipated loss of supply can have a devastating effect on our economy and the economy of our allies, and so in a no-margin environment people who wish us harm are essentially empowered. Our ability to pursue our national interests are also inhibited. If our economy and the global economy is essentially looking over its shoulder at every moment, our ability to have significant coalitions like that we brought together around Iran, I think, would be disabled. Our ability to go to our allies and say, ''Listen, we need you to stick with us; sanctions only work if, in fact, they are broadly applied and we can now give you confidence that this is not going to cause you economic harm at home,'' we were able to say that because of domestic production. Lifting the ban would only strengthen our hand.

And so while it is impossible to precisely delineate the prospective foreign policy of our energy abundance, I don't think it is exaggerated to say that our ability to fortify the global energy market will neutralize a myriad of threats and it will increase our options and strengthen our hands across the globe. And I thank you for the opportunity to be here.

[The prepared statement of Mr. Grumet follows:]

BIPARTISAN POLICY CENTER

Written Testimony
Jason Grumet
President, Bipartisan Policy Center

Before the United States House of Representatives Committee on Foreign Affairs
Subcommittee on Terrorism, Nonproliferation, and Trade
April 14, 2015

Chairman Poe, Ranking Member Keating, and members of the subcommittee, thank you for inviting me to testify today on the question of whether to lift restrictions on U.S. crude oil exports. I am the president of the Bipartisan Policy Center (BPC), which I founded in 2007 with former Senate Majority Leaders Howard Baker, Tom Daschle, Bob Dole, and George Mitchell. BPC is a Washington-based think tank develops and advocates for pragmatic, politically viable solutions to some of the nation's most complex challenges. BPC has ongoing projects in energy, national and homeland security, health care, immigration, economic opportunity and the federal budget.

My testimony today will address several core ideas:

1) The current restrictions on exporting crude oil are an anachronism. Forged in a bygone era of vulnerability, this policy is now inhibiting our ability to capitalize on America's energy strength.

2) Lifting these market barriers will strengthen our domestic economy and protect consumers. While gasoline prices are influenced by a myriad of factors, adding a reliable supply of crude oil to the global market will exert downward pressure on prices and protect US consumers from global supply disruptions.

3) The export ban is a form of resource nationalism that undermines our nation's fundamental commitment to efficient markets and our ability to promote free and fair trade.

4) By keeping US resources and market power on the sidelines, the ban empowers our adversaries to use energy as a weapon and diminishes our ability to pursue a myriad of policy and security interests.

5) Congress should move to lift these restrictions in a deliberate manner that is cognizant of the impact on those refiners that have come to rely on lower domestic crude prices.

6) We must all continue to explore the implications of this policy change and our remarkable energy abundance on a host of other complex policies from the Strategic Petroleum Reserve to the Jones Act to the Renewable Fuels Standard.

7) Of late many energy related issues have become subsumed as proxies in the critical and unfortunately polarized debate over climate change. The Bipartisan Policy Center believes that additional action is necessary to effectively address climate change. However, perpetuating inefficient markets through trade restrictions in hopes of somehow reducing global reliance on fossil fuels is not an effective climate change strategy and if anything will result in increased global emissions. Moreover, Congress in the coming months must engage in serious debates over an array of related issues from expanded oil and gas development on the OCS to ensuring safe and environmentally responsible drilling practices to reducing fugitive methane emissions to the siting of critical oil and gas infrastructure and the safety of oil transport by rail, to name a few. Lifting the oil export ban is of significant importance to our economy and must be decided on its own merits.

Overview

BPC believes that Congress and the Administration should take further steps to lift restrictions on U.S. crude oil exports. These restrictions are outdated market barriers that, left unaddressed, will undermine domestic production and our economic recovery. While the political debate will inevitably come down assertions about price of a gallon of gasoline, this issue is fundamentally a commercial dispute between oil producers who will benefit from selling their product in a competitive global market and refiners who rely on lower domestic crude oil prices (relative to international prices) to maintain profitably.

In general, lifting the ban will increase U.S. production. While no one can confidently predict the price impact of adding 1-2% of additional crude to the global market, the basic dynamics of supply and demand should give us all high confidence that increasing supply will ultimately lower the costs of crude and gasoline, and more importantly reduce the vulnerability of the global market to disruptions leading to price spikes. From a foreign policy and international security vantage point, erasing this protectionist policy sends a clear signal in favor of free trade and demonstrates that the United States is doing our part to strengthen global energy markets. By contributing to the pool of global spare capacity, we strengthen our leverage to restrain Iran's nuclear ambitions and diminish the ability of others who seek to manipulate energy supplies for their own geopolitical gain.

U.S. Oil Production – A new Reality

Two weeks ago, the U.S. Energy Information Administration (EIA) announced that the growth in U.S. crude oil production in 2014 was the highest in more than 100 years. Production increased by 1.2 million barrels per day compared to 2013—a percentage increase of over 16 percent— with most of the additional production coming from tight oil plays in North Dakota, Texas and

New Mexico. Domestic crude oil production has increased every year since 2009, after roughly two decades of declining production.[1] With domestic petroleum production having increased over 35 percent since 2009, the United States now accounts for approximately 14 percent of the total global oil supply, and is once again the largest producer of petroleum liquid fuels in the world.[2,3]

Just a few years ago, the United States was resigned to an inexorable decline in domestic oil production and increasing dependence of foreign sources of supply. The past several years have brought about a dramatic reversal. Horizontal drilling and hydraulic fracturing technologies have been applied not only to natural gas production from shale, but also to crude oil production from shale formations. The results have been surprising and spectacular. Domestic oil production has increased sharply and, most analysts believe, will continue to do so. Over the remainder of this decade, the United States is projected to increase its domestic crude oil production from 8.7 million barrels per day in 2014[4] to 9.55 million barrels per day in 2020[5]—a level not seen since 1970.[6]

At the same time, changing demographics and consumer preferences in the United States, along with ambitious new fuel economy standards and investments in energy efficiency, has led to flattening domestic demand for petroleum products. From 1983 to 2005, U.S. petroleum consumption grew by more than 35 percent, peaking at 20.8 million barrels per day in 2005.[7] From 2005 to 2014, however, the United States reduced its petroleum consumption by over 8 percent, to 19.0 million barrels per day.[8] EIA estimates that U.S. petroleum consumption will remain below 20 million barrels per day through the year 2040.[9]

In 2013, the BPC's Strategic Energy Policy Initiative issued a major report that declared unequivocally: "The state of U.S. domestic energy sectors, energy productivity, and energy security is the best it has been in many decades." This statement is even truer today than it was two years ago. It is time to embrace America's energy abundance and lift the 40-year old ban on U.S. crude oil exports.

The Mismatch between U.S. Crude Oil Production and Refining Capacity

Over the past several decades, U.S. refiners have invested tens of billions of dollars increasing capacity to refine heavier, high-sulfur "sour" crudes like those imported from Saudi Arabia, Canada, Mexico, and Venezuela. The recent increase in U.S. oil production primarily consists of light sweet crude. In response, Imports of light sweet crude to the Gulf Coast have fallen to almost zero, while light sweet crude imports to East Coast refiners have fallen by over 70 percent since 2010.[10] At current rates of production, domestic production of light sweet crude will outstrip our current domestic refining capacity.

Last week, EIA released a new report outlining possible approaches for processing the increased domestic production. The report examines a range of options, such as expanding domestic refinery capacity to process light sweet crude oil, or blending of additional light sweet crude and heavier oil. However, there are trade-offs with all of these approaches. Options that require little capital investment are limited and could result in operational inefficiencies at refineries. Options requiring major capital investments face a range of market risks. It is important to emphasize that lifting the export ban does not obligate anyone to export domestically produced crude. Our goal should be to enable the market to determine the optimal increase in domestic refining capacity and export. The current uncertainty in U.S. policy directions precludes critical infrastructure investment undermining producers, refiners and consumers alike.

Economic Impacts of Lifting the Crude Oil Export Ban

A key question for policymakers and voters is whether lifting restrictions on crude oil exports, will meaningfully affect domestic gasoline prices. In short, the answer is no. While one cannot eliminate the possibility of minor, localized price impacts while the markets recalibrate, the price of U.S. gasoline is driven by the global price of oil and elimination of the export ban will exert downward pressure on the global oil price.

It is understandable that some assume that refiners receiving below-market crude oil will "pass on" these savings to consumers. However, this is not how competitive markets function. Refiners appropriately seek the highest price for their product capturing any "windfall" from lower feedstock costs for shareholders. The U. S. has long been an exporter of refined petroleum products. As noted above, exports of refined petroleum products are not restricted under the Export Administration Regulations. Since 2001, exports of refined petroleum products, including gasoline, have increased dramatically, rising by over 300 percent.[11] Because gasoline and other refined products are traded internationally, prices in the United States for these refined products reflect international crude and refined product prices, not domestic crude oil prices. As EIA noted in its October 2014 report, "Gasoline is a globally traded commodity and, as a result, prices and changes in prices are highly correlated across global spot markets."[12]

Over the past year, a number of studies—including analyses from IHS and Columbia University—have attempted to quantify the potential economic impacts of lifting the crude oil export ban. These studies point to the possibility that without an international market for domestic crude oil, prices may be depressed to the point where upstream investment and production will be curtailed. In contrast, economic fundamentals, as described in the EIA report, point to a number of potential benefits of lifting the ban.

For instance, IHS found that over the period 2016-2030, U.S. crude oil production would be increased somewhere between 1.2 and 2.3 million barrels per day, compared to a scenario where exports are not allowed. With open exports, U.S. gasoline prices would fall 8-12 cents per gallon during this time.[13] The Columbia University study found similar results for the 2015-2025 period: lifting the ban would increase U.S. crude oil production by 0-1.2 million barrels per day and would decrease U.S. gasoline prices by 0-12 cents per gallon.[14]

The greater economic benefit from lifting the export ban is likely to come in the form of avoided harm. Until recently, the U.S. and global economies were highly vulnerable to a global oil disruption. Whether caused by accident or intentional malice, the loss of just a few percent of global production would send prices skyrocketing and the anticipation of this possibility or "risk premium" was a force in driving gasoline over $4/gallon. Increased U.S. production in recent years has contributed to a far more resilient global market place that is reflected in lower global prices. Lifting the export ban will further encourage this dynamic. As Adam Sieminski, Administrator for the Energy Information Administration, noted at a 2013 BPC event, "Two million barrels a day more production in the U.S. means, in a sense, two million barrels a day more spare capacity around the world and EIA has shown … that there is a very direct relationship between spare capacity and prices. And higher global spare capacity is almost always associated with lower and more stable pricing."

Geopolitical Impacts of Lifting the Crude Oil Export Ban

U.S. policy, both foreign and domestic, has operated under an assumption of energy scarcity for the past three decades. Today, the rules of U.S. diplomacy are being rewritten for a future less dependent on foreign oil, with significant implications for the country's strategic posture and relationships with trading partners and allies alike.

On the broad issue of trade policy, the U.S. has righteously decried the "resource nationalism" and "protectionism" that have long hindered global energy markets. Until recently, our four-decade ban on oil exports was a quaint policy aberration. While hypocritical in theory, it had no material impact as no one imagined the U.S. would ever have substantial excess capacity to trade in the global market. Happily, times have changed. A decision by Congress to perpetuate this exception now that it matters would undermine U.S. credibility in challenging trade restrictions and promoting open markets.

Increased U.S. supplies, combined with growing international production and the potential transfer of new extraction technologies, are already having ramifications for the Organization of Petroleum Exporting Countries (OPEC). Over time, it has become increasingly difficult for OPEC to make cohesive, strategic decisions, in part because its members have differing goals and needs. Many OPEC nations rely heavily on oil revenues to support their governments and to keep their populations satisfied, while others are unable to meet their production targets due to political, technical, or geological realities. Declining oil prices over the past several months have exacerbated differences among OPEC members, and numerous energy market analysts and economists, including Alan Greenspan, believe that OPEC has "lost its clout" as a result of the marked increase in U.S. oil production. Without question, OPEC's declining influence allows more flexibility for the United States to pursue its foreign policy goals. And allowing U.S. exports into the marker decreases the sway of other global oil exporters including Russia, Venezuela, and Iran.

Increased supplies of U.S. oil have helped to balance international oil markets in the face of substantial unrest in oil producing regions, and have also enabled the successful implementation of Iranian sanctions without creating additional market instabilities. Absent spare capacity in the global oil market, any action that creates a supply disruption can have a devastating effect on the U.S. and global economy. In a "no-margin" environment, those who wish to do the U.S. harm are empowered. Moreover, our ability to pursue critical national interests are inhibited if the U.S. economy and economic interests of our allies are highly vulnerable to reductions in global supply.

While it is impossible to precisely delineate the prospective foreign policy benefits of the U.S. energy abundance, it is not exaggerated to assert that our ability to fortify the global oil market neutralizes a myriad of potential threats while increasing our options and strengthening our hand across the globe.

The Path Forward

Over the past decade, technology innovations have unlocked a vast domestic energy resource. In combination with great strides in efficiency, our energy future is now defined by strength, abundance and opportunity. However, our ability to secure the promise of abundance is being hindered by a framework that was designed for a much bleaker reality. Our nation has repaired a number of these provisions. We repealed the Fuel Use Act, adopted at around the same time as the export ban, which precluded the use of natural gas in power plants. We recently reassessed our approach to exports of liquefied natural gas (LNG) arriving at the right spot of expedited export approvals after serious debate and analysis. It is now time to align the framework governing oil exports with current economic, technological and geopolitical realities.

As Congress considers lifting the export restrictions, it must also grapple with the implications of our remarkable energy abundance on a host of related policies from the Strategic Petroleum Reserve to the Jones Act to the Renewable Fuels Standard. All are affected by the dramatic changes in domestic energy production and all will benefit from reexamination in the coming years. Of late many energy related issues have become subsumed as proxies in the critical and unfortunately polarized debate over climate change. The Bipartisan Policy Center believes that additional action is necessary to effectively address climate change. However, perpetuating inefficient markets through trade restrictions in hopes of somehow reducing global reliance on fossil fuels is not an effective climate change strategy and if anything will result in increased global emissions. In the coming months, Congress must also engage in serious debates over an array of related issues from expanded oil and gas development on the Outer Continental Shelf to ensuring safe and environmentally responsible drilling practices to reducing fugitive methane emissions to the siting of critical oil and gas infrastructure and the safety of oil transport by rail to name a few. However, lifting the oil export ban is of significant importance to our economy and must be decided on its own merits.

In closing, while BPC believes that the benefits of lifting the export ban greatly outweigh the costs, there are costs, particularly to a small number of domestic refineries that may not be able to sustain current operations in a fully competitive marketplace. We hope that Congress will be receptive to suggestions that minimize these disruptions during the necessary transition to a more competitive and efficient market.

---

[1] U.S. Energy Information Administration, "U.S. oil production growth in 2014 was largest in more than 100 years," *Today in Energy*, March 30, 2015. Available at: http://www.eia.gov/todayinenergy/detail.cfm?id=20572.

[2] In 2013 (the most recent year for which EIA presents international data), the U.S. oil supply—including crude oil, lease condensate, natural gas plant liquids, other liquids, and refinery processing gain—was about 12.34 million barrels per day out of a total 90.88 million barrels per day for the world. See: U.S. Energy Information Administration, *International Energy Statistics*. Available at: http://www.eia.gov/cfapps/ipdbproject/iedindex3.cfm?tid=5&pid=53&aid=1&cid=ww,US,&syid=2009&eyid=2013 &unit=TBPD.

[3] U.S. Energy Information Administration, "U.S. remained world's largest producer of petroleum and natural gas hydrocarbons in 2014," *Today in Energy*, April 7, 2015. Available at: http://www.eia.gov/todayinenergy/detail.cfm?id=20692.

[4] U.S. Energy Information Administration, "Crude Oil Production," March 30, 2015. Available at: http://www.eia.gov/dnav/pet/pet_crd_crpdn_adc_mbblpd_a.htm.

[5] U.S. Energy Information Administration, "Table A11. Petroleum and other liquids supply and disposition," *Annual Energy Outlook 2014*. Available at: http://www.eia.gov/forecasts/aeo/pdf/tbla11.pdf.

[6] U.S. Energy Information Administration, "Crude Oil Production," March 30, 2015. Available at: http://www.eia.gov/dnav/pet/pet_crd_crpdn_adc_mbblpd_a.htm.

[7] U.S. Energy Information Administration, "Table 5.1a. Petroleum and Other Liquids Overview, Selected Years, 1949-2011," *Annual Energy Review 2011*, September 27, 2012. Available at: http://www.eia.gov/totalenergy/data/annual/pdf/sec5_6.pdf.

[8] U.S. Energy Information Administration, "Table 3.1 Petroleum Overview," *Monthly Energy Review*, March 26, 2015. Available at: http://www.eia.gov/totalenergy/data/monthly/pdf/sec3_3.pdf.

[9] U.S. Energy Information Administration, "Table A11. Petroleum and other liquids supply and disposition," *Annual Energy Outlook 2014*. Available at: http://www.eia.gov/forecasts/aeo/pdf/tbla11.pdf.

[10] U.S. Energy Information Administration, *Crude Import Trucking Tool*. Available at: http://www.eia.gov/beta/petroleum/imports/browser/#/?vs=PET_IMPORTS.WORLD-US-ALL.A.

[11] U.S. Energy Information Administration, "Exports by Destination," March 30, 2015. Available at: http://www.eia.gov/dnav/pet/pet_move_expc_a_EPP0_EEX_mbblpd_a.htm.

[12] U.S. Energy Information Administration, *What Drives U.S. Gasoline Prices?*, October 2014. Available at: http://www.eia.gov/analysis/studies/gasoline/pdf/gasolinepricestudy.pdf.

[13] IHS, *US Crude Oil Export Decision: Assessing the impact of the export ban and free trade on the US economy*, May 2014. Available at: https://www.ihs.com/info/0514/crude-oil.html.

[14] Jason Bordoff and Trevor Houser, *Navigating the U.S. Oil Export Debate*, January 2015. Available at: http://energypolicy.columbia.edu/sites/default/files/energy/Navigating%20the%20US%20Oil%20Export%20Debat e_January%202015.pdf.

Mr. POE. Thank you.

Ms. Rosenberg?

STATEMENT OF MS. ELIZABETH ROSENBERG, DIRECTOR, ENERGY, ECONOMICS, AND SECURITY PROGRAM, CENTER FOR A NEW AMERICAN SECURITY

Ms. ROSENBERG. Thank you, Chairman Poe, Ranking Member Keating, members of the committee, for the opportunity to testify today on the U.S. crude oil export ban. Recent dramatic increases in U.S. energy production have reshaped our oil industry, our industrial output and many of our global trading relationships, as my co-panelists and the prior testifiers already indicated. The oil boom has improved our GDP and balance of trade and meaningfully advanced our energy and national security. These benefits however will be clipped if policymakers do not change 1970s era crude export policies that prevent U.S. oil from moving to markets overseas.

In today's abundant oil market, supply conditions with a problematic mismatch between increasing new volumes of domestic light oil and a refining industry geared toward heavier oil, having export restrictions does not make sense. They prevent U.S. producers from accessing international buyers able to process more light crude and who will pay international benchmark prices. They depress domestic prices and distort the market. And in turn, this constrains the growth potential for domestic producers and our economy more broadly. Only a subset of American refiners benefit from the depressed domestic oil prices and they do not pass on cost savings to consumers as gasoline prices are largely set by global benchmarks.

Removing the oil export ban while promoting responsible production and energy efficiency will help to alleviate energy market distortions, and improve productivity, natural resource stewardship and economic performance. It will stimulate energy production growth which will decrease domestic gasoline prices and expand GDP.

Strengthening our economy, the engine of our national security, strengthens the United States to lead on international economic, strategic and defense matters. Lifting the ban will also support our foreign partners and our interests abroad. More U.S. crude shipped overseas will diversify the global supply pool and allow our trading counterparts abroad to achieve a better mix of imported energy commodities. This will enhance market efficiencies and lower costs for consumers.

These factors make the United States a more important trading partner for economies abroad and therefore expand U.S. leverage in trade negotiations and in the conduct of our foreign affairs. At a critical moment in the evolution of trade negotiations with Atlantic and Pacific partners, the United States should affirm a commitment to free trade and energy and expectation that trading partners will adopt similar commitments. Additionally, open energy trade is in line with U.S.-WTO commitments and will be indispensable in winning potential future natural resource trading disputes.

Another important benefit of lifting the oil export ban is the contribution it will make to energy security. When more of the oil supply pool comes from stable producers such as producers in the

United States, the overall market is more stable. U.S. crude will be shipped by fewer maritime hot spots and choke points such as the Straits of Hormuz and the South and East China Seas.

Particularly in times of market crisis, the unrestricted ability of U.S. producers to export will make them more responsive to market signals and better able to adapt quickly. This contributes to market conditions that can quickly resolve and possibly even deter actions by foreign producers to use oil as a strategic weapon. Lifting the export ban will also give the United States more flexibility to sustain and expand energy sanctions in the future. Notwithstanding the potential for a successful nuclear deal with Iran, this is important as a contingency measure, at a minimum.

Allies of the United States, many of whom reluctantly participated in energy sanctions in the past, may prove unwilling to participate in future sanctions unless the United States makes a proactive effort to stimulate alternative oil supplies and keep the market balanced. If the United States cannot convince allies to join on energy sanctions against adversaries in the future, the threat of new sanctions will not be credible and their effect will not be forceful.

Washington has a unique window of opportunity to harvest dividends from abundant domestic energy. Policymakers should lift the oil export ban and promote responsible energy production to promote economic growth and allow the United States to reap the geopolitical advantages of having a larger and more flexible role in the global oil market.

Thank you for the opportunity to testify and I look forward to answering your questions.

[The prepared statement of Ms. Rosenberg follows:]

CONGRESSIONAL TESTIMONY

The Crude Oil Export Ban: Helpful or Hurtful?
Prepared Statement of Elizabeth Rosenberg

Center for a New American Security

April 14, 2015
Testimony before the U.S. House Committee on Foreign Affairs, Subcommittee on Terrorism, Nonproliferation, and Trade
Prepared Statement of Elizabeth Rosenberg
Senior Fellow and Director, Energy, Economics and Security Program
Center for a New American Security

Chairman Poe, Ranking Member Keating and members of the committee, thank you very much for the opportunity to testify today on the U.S. crude oil export ban. I will focus my remarks on the national security and foreign policy implications of exporting domestic crude oil.

Recent dramatic increases in U.S. energy production have reshaped our oil industry, industrial output and many of our global trading relationships. The United States has expanded oil production by 88% since 2008,[1] cut net oil imports by 31% since this time,[2] and according to the International Energy Agency, will account for the greatest source of global oil supply growth through 2020.[3] The energy revolution has strengthened GDP and balance of trade conditions over the last several years.[4] Additionally, it has helped to stabilize the global energy market during a period of record, sustained supply disruption. By strengthening our global trading position and our economy, the engine of our national security, the energy revolution has meaningfully advanced our security and the ability of the United States to lead on foreign affairs.

Going forward, our remarkably productive, innovative and resilient energy sector can deliver even further benefits to U.S. economic and national security. However, these benefits will be clipped if policymakers do not change antiquated crude export policies that prevent U.S. oil from moving to markets overseas. In a domestic market awash with oil, keeping 1970s-era export restrictions in place discriminates against U.S. producers and threatens investment in new supply, thereby jeopardizing economic, security, and trade gains from the energy boom. Policymakers should lift the oil export ban to bring export policy in line with present market

---

[1] Energy Information Administration, "Weekly Supply Estimates,"
http://www.eia.gov/dnav/pet/pet_sum_sndw_dcus_nus_w.htm.
[2] Energy Information Administration, "Weekly Imports & Exports"
http://www.eia.gov/dnav/pet/pet_move_wkly_dc_NUS-Z00_mbblpd_w.htm.
[3] Lejla Alic et al., "Oil Medium-Term Market Report 2015: Market Analysis and Forecasts to 2020," (International Energy Agency, 2015), 41.
[4] John W. Larson et al., "America's New Energy Future: The Unconventional Oil and Gas Revolution and the US Economy Volume 3: A Manufacturing Renaissance- Main Report," (IHS, September 2013).

CONGRESSIONAL TESTIMONY

The Crude Oil Export Ban: Helpful or Hurtful?
Prepared Statement of Elizabeth Rosenberg

Center for a
New American
Security

circumstances, to promote free trade and responsible growth in the sector, and to reap the geopolitical advantages of having a larger and more flexible role in the global oil market. As responsible stewards of our natural resources, policymakers should couple this change with the promotion of energy efficiency and of low-carbon fuel sources at home and abroad.

U.S. Prohibitions on Oil Export No Longer Make Sense

The current oil export restrictions were established four decades ago on the heels of a series of energy price controls and supply allocations. In light of the Arab OPEC oil embargo of 1973 and subsequent large oil price increases imposed by OPEC countries, the legislatively enacted ban on crude export was intended to promote energy and economic security. Slight modifications have been made to the export prohibitions over time, by both Democratic and Republican presidents, allowing a few exceptions, and price controls and supply allocations were removed in the early 1980s. The crude oil ban, however, was not removed, a circumstance which did not have much practical effect until recently, given the heavy dependence of U.S. consumers on imported oil for so many decades. However in today's abundant oil market supply conditions, with a problematic mismatch between the increasing new volumes of light quality oil produced in the United States and a refining industry geared toward heavier crude, these rules do not make sense.

Export restrictions create barriers for domestic producers trying to sell their oil and distortions in the market. Many domestic producers sell their light quality crude at a discount because of its abundance relative to demand, the unsuitability of processing too much of it at domestic refineries oriented towards heavier crude, and infrastructure bottlenecks that make the journey to market more difficult and expensive. Critically, however, they have to sell their light quality crude at a discount because U.S. producers are restricted from exporting this crude abroad. The export ban prevents them from accessing international buyers better able to process more light crude and who will pay international benchmark prices for this oil. In this situation, domestic oil producers see a check on their growth potential. It is only a subset of the U.S. manufacturing sector that benefits from this market distortion. Notably, refiners in this position do not pass on their cost savings to U.S. consumers, as pump prices are largely determined by global benchmarks.[5]

So far, U.S. producers have sold their expanding crude volumes to U.S. refineries, or exported them using the limited exceptions allowed under current restrictions. These include sending oil to Canada, exporting condensate or through narrow swap arrangements. However, the point at

---

[5] Energy Information Administration, "What Drives U.S. Gasoline Prices?" (Energy Information Administration, October 2014), 7.

which producers may exhaust these options may not be far off, and may already be occurring at certain times of the year in certain circumstances, such as periods of refinery maintenance when demand for oil diminishes. Historically high levels of crude building up in inventories makes the outlook for domestic oil market saturation even more concerning. With limited relief valves for the abundant crude in the U.S. market, it also hems in the potential for domestic producers to achieve the 500,000 or more barrel per day increase in production this year anticipated by the U.S. Energy Information Administration and other independent analysts.[6]

The present crude export policy that strands light crude in the U.S. market is hardly an optimal arrangement for productivity, efficiency and economic growth. A more beneficial policy for promoting market stability, growth and security is a policy that would encourage responsible U.S. production of oil, efficient, open markets and a larger share of global oil supply from reliable producers, such as the United States. A more permissive, even encouraging, oil export policy would support these goals by allowing U.S. producers to fetch premium prices abroad. Lifting crude export restrictions makes sense even as lower oil prices slow investment and drilling in the United States, and domestic refiners consider expanding their capacity to absorb more light oil. These factors may delay the point at which the U.S. market is totally saturated with crude and the export restrictions stall out domestic oil production growth. However, responsible policy should intervene far before the oil market reaches such dire conditions.

National Security Implications of a New Oil Export Policy

Strengthening our Economy

There are a variety of economic benefits associated with lifting U.S. crude oil export restrictions that will directly benefit our national security. A variety of government and independent studies suggest that lifting the oil export ban would result in an increase in U.S. oil production, a decrease in domestic refined product prices, and growth in GDP. Oil output could rise between 110,000 barrels per day and 2.8 million barrels per day by 2020, according to these studies, with a corresponding bump in economic growth and benefit for the U.S. balance of trade.[7]

---

[6] Energy Information Administration, "Short-Term Energy Outlook," (Energy Information Administration, April 7, 2015), http://www.eia.gov/forecasts/steo/index.cfm.

[7] On the lower end of the spectrum of estimates for increases in domestic oil production, an industry-commissioned study by consultant ICF International estimated an oil production increase by approximately 110,000 to 500,000 barrels per day by 2020. (Harry Vidas et al., "The Impacts of U.S. Crude Oil Exports on Domestic Crude Production, GDP, Employment, Trade and Consumer Costs," (ICF International, March 2014), 10.) A study by NERA Economic Consulting estimated that oil production would increase by 1.3 million barrels per day to 2.8 million barrels per day by 2020 with the ban lifted in 2015. (Robert Baron et al., "Economic Benefits of Lifting the Crude Oil Export Ban,"

CONGRESSIONAL
TESTIMONY

**The Crude Oil Export Ban:
Helpful or Hurtful?**
Prepared Statement of Elizabeth Rosenberg

Center for a
New American
Security

Strengthening our economy, including reducing our international indebtedness, strengthens the stature and ability of the United States to lead on international economic, strategic and defense matters. In an era of budget austerity, war fatigue, proliferating security challenges, and the expanding use of economic sanctions, a strong U.S. economy expands policy options from some of the more conventional diplomatic and military choices. It creates an opportunity to hone smarter and more creative tools to advance our national interests in the international arena. Additionally, a more favorable trade balance also liberates the United States to consider international trade policies and international lending that could be constrained, including by some of our key economic partners, such as China, in a scenario of greater U.S. indebtedness.

In addition to providing an economic boost at home, lifting the oil ban will beneficially accrue economic yields to our foreign trading partners. A U.S. energy export policy that allows the free flow of all energy commodities—including crude oil and not just condensate and refined products—will enable U.S. and foreign trading partners to optimize trade in various kinds of energy commodities, depending on seasonal and regional demands. The greater diversity in energy commodity trading relationships will support greater energy market efficiencies, lower costs for consumers, more limited risks and greater economic growth. These factors can make economic planning more dynamic, easier and reliable for policy leaders abroad, and in the United States. Additionally, these factors can make the United States a more important trading partner for more energy consumers abroad, a circumstance which will expand the soft power leverage of the United States in international strategic relationships.

Promoting Open Markets

Lifting the restrictions on export of domestic crude will allow U.S. policy leaders to set the right anti-protectionist tone on trade in the international arena and reap economic and strategic benefits from an open energy market system. At a dynamic time in global energy trade and a critical moment in the evolution of U.S. free trading terms with partners across the Atlantic and the Pacific, U.S. policy leaders have a unique opportunity to send a strong message on a commitment to open markets by lifting restrictions on oil export. In turn, this will affirm the expectation that key trading partners will adopt similar commitments on energy trade. Having more open energy trade is in line with U.S. World Trade Organization commitments, and will be

Prepared for The Brookings Institution, (NERA Economic Consulting, September 2014), 138, 139, 146, and 147.)
According to a study by IHS, total U.S. crude oil output is expected to rise between 1.08 and 1.99 million barrels per day by 2020. (Mohsen Bonakdarpour et al., "US Crude Oil Export Decision: Assessing the Impact of the export ban and free trade on the US economy," (IHS, May 2014)).

CONGRESSIONAL TESTIMONY

The Crude Oil Export Ban: Helpful or Hurtful?
Prepared Statement of Elizabeth Rosenberg

Center for a
New American
Security

indispensible in winning potential, future natural resource trading disputes that may arise with other countries.

Making a firm commitment to open energy trade will also help the United States to influence trading policy priorities in other countries, such as those in East Asia. In that region, key decisions will be made over the coming years about the nature of international energy commodity market participation that will have a direct bearing on the U.S. economy. Furthermore, the United States will be more credible in encouraging developing economies, such as China and India, to join Organisation for Economic Co-operation and Development (OECD) economies as proponents of free trade and responsible stakeholders in collective energy crisis management if Washington actively shuns protectionism.

Enhancing Market Stability

By encouraging the expanded production of U.S. crude, a result of lifting the oil export ban, policymakers will be facilitating the greater flow of oil from a reliable, secure producer to the global market. When more of the oil supply pool comes from producers that do not suffer threats from political instability or imminent danger to critical energy infrastructure or supply lanes, the overall market is more stable. Additionally, U.S. crude will be shipped to consumers overseas via fewer maritime hot spots and choke points, such as the Straits of Hormuz and the South and East China Seas. Major consumers in East Asia, for example, are highly vulnerable to supply disruptions that could occur in these areas, and are vulnerable to destabilizing conflict in the Middle East, from which a majority of their oil imports derive.[8]

Particularly in times of oil market crisis that originate outside the United States, the unrestricted ability of U.S producers to export oil will make them more responsive to market signals, and better able to quickly adapt to the needs of oil purchasers. This will contribute to market conditions that will more quickly resolve, and possibly even deter, actions by some producers to use oil as a strategic weapon. This in turn will give U.S. policymakers more options for potentially using the Strategic Petroleum Reserve in innovative and proactive ways, including counteracting hostile attempts by foreign producers to manipulate consumers or prices. If policies within reach, such as a loosening of the oil export ban, can lessen the potential for U.S. consumers to be held hostage to coercive market power, they should be very seriously considered and if at all possible, adopted.

Supporting Our Allies

---

[8] Lejla Alic et al., "Oil Medium-Term Market Report 2015: Market Analysis and Forecasts to 2020," 86.

CONGRESSIONAL TESTIMONY

The Crude Oil Export Ban: Helpful or Hurtful?
Prepared Statement of Elizabeth Rosenberg

Center for a New American Security

For our European allies, the presence of more U.S. oil in the market will offer more supply options. This will mean that European consumers look less to Russia, from which they receive roughly 30% of their oil supplies[9] and which has a history of coercive energy supply policies. When Russia has more competition for supplying European demand it will have to work harder to play a role in the market.

A fundamental pillar in the current U.S. policy regarding Ukraine and Russia's destabilizing role there involves degrading Russia's ability to compete in the global oil market, even while that may cause a moderate economic effect on the U.S. and European economies. A liberalization of U.S. oil export policy will have the effect of reinforcing the pressure on Russia's energy sector and is certainly in line with key U.S. national security goals. It will also constitute an important strategic act of support for allies in Europe, who are more threatened by Russian regional destabilization. When our closest allies are stronger, the United States is more secure and better able to bolster and lead multilateral security initiatives to counter global threats.

For East Asian partners, more U.S. oil supply in the market would give them new opportunities to diversify away from increasingly unstable Gulf and Russian oil supplies. In addition to boosting supply security, such diversification will yield greater market efficiencies and will contribute to lower prices. This will be true for all Asian nations, including both our treaty allies in Northeast Asia and China. Policies that confer mutual benefit on the United States and the group of East Asian nations facing off as regional competitors should be priorities for the United States. They may help to deter strategic intra-regional competition by increasing the shared incentives for stable, efficient market activity. Enhancing stability in this neighborhood is directly in line with the United States' policy of rebalance to Asia, and will benefit our country and all others that see their own stability tied to stability of this burgeoning region. Putting in place policies that can contribute, even if modestly, to enhancing regional stability will cultivate the influence of the United States in Asia and beyond.

Expanding Sanctions Leverage

One of the most important security benefits of lifting the crude export ban is the additional flexibility and leverage it will give to the United States to sustain and expand energy sanctions in the future. Policymakers in the United States have looked increasingly to energy sanctions over the last several years as a policy instrument to isolate and coerce adversaries. Economic sanctions

---

[9] European Commission, EU Crude Oil Imports Statistics, http://ec.europa.eu/energy/en/statistics/eu-crude-oil-imports.

CONGRESSIONAL TESTIMONY

The Crude Oil Export Ban: Helpful or Hurtful?
Prepared Statement of Elizabeth Rosenberg

Center for a
New American
Security

that reduced Iran's oil exports by almost 60% from approximately 2.5 million barrels per day in 2012 to 1.1 million barrels per day now[10] are credited with bringing Iran to the negotiating table over its nuclear enrichment program. Particularly in light of historically high oil supply disruptions globally, the international community would not have been able to sustain these sanctions, and cope with the oil price increases they would have caused, were it not for massive increases in alternative oil supplies. The United States added about 1 million barrels per day annually over the last several years, and Saudi Arabia also turned up production to balance the market.[11] In addition to targeting Iran's energy sector, the United States and the European Union have also imposed sanctions on Russia to handicap its energy sector as part of the broader Ukraine policy strategy.

With the announcement of framework understandings between Iran and the P5+1 negotiators over Iran's nuclear program earlier this month, the outlines of a potential final agreement that would relieve many sanctions on Iran is taking shape. Whether the negotiators conclude a deal by their June deadline or not, policymakers will need to enhance their ability to impose tough additional energy sanctions in the future. This is critical as an element of contingency planning on Iran policy and to provide a credible threat that more oil sanctions on Iran are possible if Tehran does not cooperate with the international community. Additionally, a grim outlook on relations with Russia, and the attractiveness of the energy sanctions tool to attack other potential new security problems in the future, means that policymakers should cultivate the ability to potentially deploy energy sanctions in multiple theatres simultaneously.

The failure to prepare for the potential future imposition of more energy sanctions by stimulating alternative oil supplies may render the threat of new sanctions hollow. If adversaries do not believe that the United States and its allies have the economic and political tolerance to cope with a self-imposed oil price increase, which could occur if more sanctions pull more oil off the market, these adversaries may call a bluff. Furthermore, allies of the United States, many of whom have reluctantly gone along with energy sanctions in the past, may prove unwilling to participate in further energy sanctions unless the United States makes a serious effort to stimulate alternative oil supplies. Lifting the U.S. oil export ban will bring online additional U.S. production, and would constitute an important signal to allies, adversaries and market participants alike, that the United States is serious about the threat, or actual use, of forceful energy sanctions.

[10] David S. Cohen, Under Secretary for Terrorism and Financial Intelligence, "Written Testimony of David S. Cohen," Statement to the United States Committee on Foreign Relations, U.S. Senate, January 21, 2015, 5.
[11] Lejla Alic et al., "Oil Medium-Term Market Report 2015: Market Analysis and Forecasts to 2020," 61.

CONGRESSIONAL TESTIMONY

The Crude Oil Export Ban: Helpful or Hurtful?
Prepared Statement of Elizabeth Rosenberg

Center for a
New American
Security

Conclusion

In a period of tremendous geopolitical uncertainty, and when many questions exist about the future role of the United States as a global energy player and world leader, Washington has a unique window of opportunity to strengthen domestic economic growth, oil market stability, U.S. global leadership and open trade relations. Removing the outdated, discriminatory and detrimental ban on the export of U.S. crude oil will advance these goals. It will deepen trading ties with strategic allies, including those in Europe and Northeast Asia. It will improve the economic position and energy market stability of our nation and partners abroad, and allow the U.S. to more effectively spur and lead multilateral action to counter international security threats. Taking this action should be coupled with a policy focus on responsible energy production, efficiency and the promotion of low-carbon fuels. Bringing together these measures, a new national energy policy can enhance energy and national security, and expand our ability to advance targeted foreign policy measures in the future.

Mr. POE. We now turn to Mr. Bordoff for your statement.

STATEMENT OF MR. JASON BORDOFF, FOUNDING DIRECTOR, CENTER ON GLOBAL ENERGY POLICY, COLUMBIA UNIVERSITY

Mr. BORDOFF. Chairman Poe, Congressman Keating, members of the committee, thank you for the invitation to be with you today. I would like to summarize some of the findings from a recent report I co-authored at the Columbia University Center on Global Energy Policy, copies of which you should have in front of you.

The oil export ban was originally adopted in the 1970s in response to concerns not only about oil scarcity after the Arab oil embargo, but also to prevent producers from getting around domestic price controls by exporting their oil into the global market where they could fetch a higher price. Price controls were eliminated 30 years ago, but the export restrictions remain.

U.S. oil production, as we all know, has boomed and imports have plummeted as a result. We are still going to be an importer of oil though for as far as the eye can see, most likely, so why are we even talking about exports? As we talked about this morning, the concern is the ability of domestic refiners to absorb the kind of oil that we are producing in the U.S. U.S. shale oil is very light oil, while many of our refineries have invested billions to handle heavy sour oil. You can run light oil through those refineries but it becomes increasingly economically challenging to do so. So as we have heard the price of U.S. oil may become discounted relative to the world price to incentivize domestic refiners to take it.

It is about $6 today for a variety of reasons. A lower U.S. price would in turn mean less U.S. production, lower economic activity and higher net imports. To date, U.S. refiners have made low cost adjustments where they can. We have backed out mostly the import of light oil, and we have also exported what is allowed. Exports after all are not completely banned. They are restricted. Exports are allowed, for example, to Canada, and our exports there have surged, to almost ½ million barrels a day. And we have also had a surge in the export, as you heard, of refined petroleum, which is also allowed.

As U.S. production grows, however, at some point you run out of these low cost options. The oil price crash means that the pace of U.S. supply growth is slowing down. The Energy Information Administration said yesterday production will probably decline next month, the first decline in U.S. shale oil output in 4 years. Storage is at an 85-year high.

However, the oil export issue is still relevant. First, production may rebound faster than we expect. Second, the export ban may still depress U.S. prices periodically and temporarily, for example, during refinery maintenance or in response to other outages. Third, U.S. production may be more sensitive to any price discount at today's lower levels. And then fourth, the policy process takes time. So I think it makes sense to prevent a market problem rather than wait for one to develop and then respond to it.

Now what about gasoline prices? Wouldn't lower U.S. oil prices mean lower U.S. pump prices? Well, we have talked about that already. The answer is no. This is because gasoline and diesel can be freely traded in the global market so the price is set by the world price. If U.S. crude is discounted that means refineries can buy crude more cheaply, but they still sell their product at the global price. And we saw this from 2011 to 2013 when the U.S. price was depressed not by the export restriction but by pipeline bottlenecks in Cushing. And as you heard, the Brookings Institution, Resources for the Future, Rice University, the Energy Information Administration and others, have found exports won't raise pump prices, they might slightly lower them.

We also want U.S. supply to respond to global circumstances. Consider how OPEC decided in November to let oil prices fall, forcing higher cost producers like the U.S. to cut production instead. We know shale oil can go off line very quickly compared to conventional oil, but it can also bounce back quickly too. And if the world price were to rise again to the $70s or $80s or $90s, U.S. oil supply could rebound quickly to slow that price rise to temper the impact on consumers at the pump. But that U.S. supply response may be impeded if we have to sell our crude at a discounted price.

Briefly I would add, allowing exports I think is consistent with America's longstanding commitment to free and open markets and it enhances our credibility in trade negotiations and avoids creating a potentially harmful precedent. Increased U.S. supply can also weaken the economic power, fiscal strength and geopolitical influence of other large oil producing companies and enhance U.S. diplomatic leverage in certain circumstances.

And then, finally, I want to talk about the critical issue of climate change. We need to do much more to address climate change. To the extent oil exports boosts U.S. oil supply and lowers global prices, oil use and carbon emissions will rise, but climate change, I think, is best addressed with policies targeted at that problem. Restricting trade is a very costly way to achieve modest emission reductions relative to the benefits. Many government policies may raise emissions, like achieving faster GDP growth or a deal with Iran that allows Iran to sell more oil, but may still be desirable when the benefits are weighed against the costs.

Members of the committee, thank you again for inviting me to appear here today and I look forward to answering your questions.

[The prepared statement of Mr. Bordoff follows:]

COLUMBIA|SIPA
Center on Global Energy Policy

April 14, 2015

Congressional Testimony of
Jason E. Bordoff
Founding Director, Center on Global Energy Policy, and Professor of Professional Practice in International and Public Affairs, Columbia University School of International and Public Affairs

Before the
Committee on Foreign Affairs, Subcommittee on Terrorism, Nonproliferation, and Trade
United States House of Representatives
1st Session, 114th Congress

OVERVIEW

The oil export ban was originally adopted in the 1970s in response not only to concerns about oil scarcity, but to prevent oil producers from getting around domestic price controls by selling oil into the global market for a higher price. Price controls were repealed in 1981. But even though **the original rationale for the statutory export ban was eliminated over 30 years ago, the export restrictions remain**.

US oil production has boomed over the past few years and led to sharply lower import dependence, but we are still likely to remain a net importer, consuming more than we produce. So why are we talking about exports? The concern is the **ability of domestic refiners to absorb the *kind* of oil we are producing** – US shale oil is very light oil, while many of our refiners have invested billions to handle heavy, sour oil. You can run light oil through a one of those refineries, but it is increasingly economically challenging to do so. If US production continues to grow, the **price of US oil may increasingly be discounted** relative to the world price to incentivize domestic refiners to take it. A lower US price would, in turn, mean **less US production, lower economic activity, and higher net imports**.

Allowing unrestricted crude exports would enable US producers to send light oil to refineries that want it elsewhere and import heavier oil to run in our refineries if it is more economic to do so.

A recent study I co-authored with Trevor Houser of Rhodium Group for the Columbia University Center on Global Energy Policy found that allowing US producers to access global crude markets would **increase US oil production**—although the magnitude and timing of these impacts are highly uncertain. Because gasoline prices are set in the global market, oil exports **would not raise pump prices**, and might even lower them slightly. Lifting oil export restrictions is also consistent with past and present **US trade policy objectives** and yields **geopolitical benefits**. To the extent exports lower oil prices and lead to higher oil use, lifting the statutory restriction would also **raise carbon emissions**. While export restrictions are **neither an appropriate nor cost-effective** way to address climate change concerns, it is critical that **more aggressive climate policy actions** be taken to reduce greenhouse gas emissions.

COLUMBIA | SIPA
Center on Global Energy Policy

CRUDE OIL EXPORT BAN BORN FROM SCARCITY CONCERNS OF THE 1970s

The 1970s shook the oil industry to the core and brought energy security to the fore of American public consciousness. Resource nationalization, the end of the dominance of the "Seven Sisters" international oil companies, the Arab oil embargo, and the revolution in Iran redrew the global energy map. These events are often credited with giving rise to concerns about oil "scarcity" that ultimately led to restrictions on the export of oil, although the seeds of the oil export ban were sown years earlier through restrictions on oil trade and oil price controls.

US oil production peaked in 1970 and began a decades-long decline. This coincided with a series of far-reaching economic measures by President Nixon, including price and wage controls. Oil exports were not an issue at first, as the price of crude within the United States was higher than international levels. After the 1973 Arab oil embargo, however, international crude prices soared, giving US producers an incentive to sell their crude abroad. To defend domestic price controls, the government introduced restrictions on exporting crude oil and refined products such as gasoline and diesel fuel. Absent export restrictions, US producers could have skirted the price controls by selling crude oil directly into the global market for a higher price.

The oil export restrictions result from a series of laws enacted in the 1970s, most notably the Energy Policy and Conservation Act of 1975. In the early 1980s, the Reagan Administration eliminated price controls and allowed refined petroleum exports, but restrictions on crude oil exports persist.

CURRENT CRUDE EXPORT REGULATIONS

The crude export laws have been modified in various ways through executive branch actions by both parties since the 1970s. Although crude oil exports are restricted, they are **not entirely banned**. The most notable exception is for exports to Canada, which rose to nearly 500,000 barrels per day in early 2015 from just 67,000 barrels per day in 2012. Swaps and exchanges are also permitted under certain circumstances, most notably with Mexico.

As mentioned, refined petroleum may also be exported, and the US has gone from being the largest net importer of petroleum products in the world in 2006 to the largest exporter today. The ability to export refined petroleum has also brought new scrutiny to the technical distinction between "crude oil" and "refined product," which is crucial to US export policy. On December 30, 2014, the Commerce Department issued a set of Frequently Asked Questions that identified factors it will consider in making this determination. At a minimum, the revised guidance made clear that lightly processed condensates (very light liquid hydrocarbons) may be exported as refined product.

IMPACTS ON US OIL PRODUCTION

The surge in US oil production has created questions about the ability of US refiners to handle this growing supply. Many US refiners on the Gulf Coast had spent billions over the past few decades to enable their plants to run optimally with heavy, high sulfur (called "sour") imported crude. New US shale oil production is light, low sulfur, or "sweet" crude.

COLUMBIA | SIPA
Center on Global Energy Policy

These existing refineries can absorb lighter oil, although doing so becomes increasingly economically challenging as processing limits are encountered. Even with additional investment, refineries optimized for heavier crudes can still be challenged by lighter oil and by the inability to fully utilize expensive downstream upgrading equipment. As a result, as refineries over time idle high cost processing equipment or incur the cost of building new capacity, they may require a discount from domestic crude producers to justify this change in their crude slate.

A discounted US oil price resulting from the export restriction would mean less investment and less US oil production, which reduces economic activity and employment in the oil sector. In our study, we found that easing the export restriction would boost US production from anywhere between zero and 1.2 million barrels per day on average through 2025.

To date, we have accommodated the boom in US light oil in three ways. First, refineries have made low-cost adjustments to absorb more. An April 2015 study from the US Energy Information Administration found that the low-cost steps were limited or mostly already being taken. Second, we have backed out imports of light oil, especially from West Africa, although the ability to back out more light oil imports to the East Coast is challenged by the higher shipping costs of Jones Act tankers to move domestic crude to East Coast refineries. And, third, we have been exporting vastly more crude oil where allowed (almost entirely to Canada) and refined petroleum.

As US production grows, however, these existing outlets become more and more limited. US domestic commercial crude storage levels just reached an 85-year high. There is uncertainty about exactly when the "point of saturation" will be reached at which US crude prices become significantly discounted relative to the world price because of domestic processing limits. Refinery consultants Turner Mason have estimated the "point of saturation" will be reached when US production reaches between 10 and 11 million barrels per day. EIA noted, however, in its new study that higher-cost options to process light oil requiring significant investment were challenged not only by costs, but also by policy uncertainty about whether current crude oil exports restrictions will be relaxed.

The exact point at which this limit is reached depends on factors including how quickly US production grows and the ability of the US refining sector to adapt to that growth. In response to the recent collapse in oil prices, US production growth has slowed dramatically and will likely soon peak for 2015 at around 9.4 million barrels per day before picking up again in 2016. EIA's long-term forecast projects US crude oil production will peak at 9.6 million barrels per day.

Although the oil price crash has slowed the growth of US oil supply, the oil export issue is still relevant, however. First, production may rebound faster than we expect; indeed, actual production has consistently surpassed EIA projections in recent years. Second, the impact of the oil export ban may increasingly be seen seasonally and temporarily as refineries shut down for maintenance or other reasons (e.g., labor strikes or fires), causing the US price to fall further below the world price as US refining demand drops. Indeed, in early 2015 and late 2013, the price of US crude became sharply dislocated from world prices for a variety of reasons, and this seasonal weakness would likely have been reduced if US producers could export to meet global demand. Third, US production levels are more sensitive to any sort of price discount at today's lower prices in the $50s or $60s per barrel than at $100 or more, given that we are much closer to the break even costs now for many shale wells. Finally, it takes the policy process some time to build consensus and change existing laws

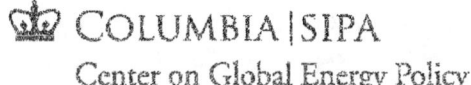 

COLUMBIA | SIPA
Center on Global Energy Policy

or regulations, so it makes sense to prevent a market problem from developing rather than wait to respond to one.

WHAT ABOUT GASOLINE PRICES?

Perhaps the key issue, substantively and politically, in the debate about whether to allow unrestrained crude exports has been the perception that such a move would push up prices at the pump for consumers. **Both economic theory and empirical evidence, however, suggest refined product prices would not rise, and may even fall slightly, if export restrictions were removed.**

Gasoline and diesel produced in the United States can be freely traded in the global market, and thus the price at the pump is determined by the world price of refined petroleum. If the US price of crude is discounted, that lowers the cost to refiners of buying crude oil to produce gasoline, diesel and other products. But there is no reason why the domestic refiners would pass those savings along to consumers. US refiners will have access to global product markets and the ability to sell gasoline and diesel abroad at prevailing global prices.

Indeed, this is exactly what's occurred over the past few years. Between 2011 and 2013, Midwest refiners paid 16 percent less, on average, per barrel of crude than East Coast refiners, thanks to infrastructure bottlenecks between the US Midcontinent and the East Coast. Refiners in the Rocky Mountain region paid 22 percent less. Yet the price of gasoline sold by Midwest and Rocky Mountain refiners was only 1 percent and 1.4 percent lower than East Coast refiners over this period respectively. Lower crude costs improved refiner profitability but did not lower prices for consumers.

The finding that unrestricted oil exports would not raise pump prices is consistent with studies by the Brookings Institution, Resources for the Future, Rice University's Baker Institute, and the Energy Information Administration, among others. To the extent lifting the oil export restrictions boosts US production relative to what it would otherwise be by allowing US producers to sell at a higher price, and to the extent that increased supply is not offset by production cuts elsewhere in the world, the increased global supply will push down gasoline prices. In our study, we estimated the reduction in gasoline prices to be between zero and 12 cents per gallon, although I would again stress the magnitude is highly uncertain and may well be small.

Unrestricted crude exports also allows US supply to respond better to global oil market signals. This consideration may be even more relevant given the nature of US shale oil and OPEC's November 2014 decision not to reduce production but to let oil prices fall, forcing higher cost producers like the US to cut production instead. Because of shale oil's very steep decline rates relative to conventional oil sources, it can go offline very quickly when producers idle rigs in response to lower prices. Indeed, the rig count has fallen in half, and US production may have stopped growing for 2015. But shale oil can bounce back very quickly too when prices rise. That means, assuming OPEC maintains its current policy, that US oil can be a new kind of "swing supply" in the global market. If the world oil price were to rise again into the $70s or $80s or beyond, **US supply could rebound quickly to slow the price rise and temper the concomitant rise in consumer pump prices**—but that US supply response may be impeded if producers have to sell at discounted prices.

COLUMBIA | SIPA
Center on Global Energy Policy

THE ENERGY SECURITY CONSEQUENCES OF ALLOWING OIL EXPORTS

Allowing unrestricted exports would make the US more resilient, not less, to supply disruptions elsewhere in the world. Greater integration into global markets would make US oil supply more responsive to international market developments, mitigating the impact on American consumers and the US economy of production losses in other countries.

Today's oil market is very different than it was during the 1970s. Then, a disruption in contracted shipments could result in a physical shortage for the buyer because of a lack of strategic or commercial stockpiles or a spot market. Today, however, the oil market has become the largest and most liquid commodity market on earth. A supply disruption anywhere raises crude prices everywhere, incentivizing both additional sources of supply and greater conservation. Interdependence means that when crude or refined product markets are disrupted, the US can mitigate supply disruptions by accessing alternative sources of supply.

Oil trade also provides economic security benefits. Broadly speaking, oil price shocks impact the US economy in three ways. First, they increase business costs and reduce real household income. Second, they put upward pressure on prices economy-wide, which can result in tighter monetary policy. Third, as long as the United States is a net oil importer, oil shocks deteriorate the country's terms of trade and can result in large temporary increases in the country's current account deficit. To the extent lifting crude export restrictions increases US production, net US oil imports will decline. This is true even though gross imports increase as more light oil is exported and more heavy oil imported than would be the case were the export restriction to remain in place. In a recent report, the White House Council of Economic Advisers (CEA) found the "resilience of the economy to international supply shocks—macroeconomic energy security—is enhanced by reducing spending on net petroleum imports and by reducing oil dependence." This is due both to the smaller terms of trade penalty from an oil price shock, and the fact that more of the increase in oil producer revenue stays within the United States.

At the same time, if lifting crude export restrictions results in a decrease in gasoline and other refined product prices, US oil demand will grow, exacerbating the impact of a given change in prices on household incomes, business expenses and overall inflation. Given the magnitude of the potential refined product price decline that would be expected, the impact on overall US oil demand would be small, however, so overall net imports would still decline.

TRADE CONSIDERATIONS

Lifting crude export restrictions is **consistent with America's longstanding commitment to free and open markets,** would **enhance US credibility** in current and future trade negotiations, and avoid creating **a precedent that could harm US trade policy objectives** down the road.

Since the founding of the postwar global trading system, the United States has been a leading proponent of open trade. For most of that time the United States was a net energy importer, so access to international energy and natural resource supplies was an important trade policy priority. The United States has also traditionally supported open international trade on the principle that it

39

COLUMBIA | SIPA
Center on Global Energy Policy

improves economic welfare both for importers and exporters. With the surprise turnaround in US oil production and trade balance, and with crude export restrictions beginning to distort trade outcomes, America's commitment to free trade principles is now being put to the test.

The US has won cases in the World Trade Organization against China and other countries when these countries tried to defend commodity export restrictions using one of the many exceptions in international trade law. Should the United States choose to maintain current crude export restrictions, it could be in the position of having to make the same arguments that it successfully defeated in these other trade disputes. The precedent established in those cases would make a US defense more challenging. Were the United States to succeed in arguing for exceptions, it would create a **precedent that could limit the ability of the US to challenge other countries' export restrictions** in the future.

Equally important is assessing the implications of maintaining US export restrictions on US credibility in other **US trade policy priorities**, such as the current negotiations with Europe over the Transatlantic Trade and Investment Partnership (TTIP). In the negotiations over the TTIP, the Europeans have argued for the inclusion of an energy chapter and the elimination of US energy export restrictions. A leaked EU document noted how maintaining export restrictions might undermine joint efforts to combat export restrictions in China and elsewhere: "Combatting resource nationalism, together vis-à-vis third countries while at the same time allowing for export restrictions to exist between us sends the wrong message to our partners and offers some of these resource-rich countries a great opportunity to interpret trade rules in a way which is detrimental to our economies."

GEOPOLITICAL CONSIDERATIONS

Increased US crude production can **weaken the economic power, fiscal strength and geopolitical influence of other large oil producing countries**. Additional supply on the market also increases competition and reduces any one country's ability to leverage its resources to gain geopolitical influence. Reducing foreign producer's oil revenue also risks negative geopolitical consequences, however, if it leads to greater instability in these regions. The magnitude of any export-driven impact is small, however, relative to recent oil market developments.

Also important for US foreign policy are the **current crude trade relationships retained and new ones created** if export restrictions are modified or lifted. If export restrictions were eased, net imports would be lower, but total gross imports and exports would be higher as refiners import crude best suited to their needs and producers export other types of crude better suited to refiners abroad. While it should theoretically make little difference where a country buys its crude from given the size and liquidity of the global market, specific bilateral trade flows can have significant geopolitical implications in practice. Beyond the direct economic gains of trade, trade generally improves bilateral relations more broadly, opens new lines of communication and reduces the odds of conflict.

Permitting exports also has the potential to **boost US diplomatic leverage** in certain circumstances, such as the future application of sanctions or pursuit of other objectives. The application of sanctions against Iran, for example, depended critically on US diplomacy to persuade

COLUMBIA | SIPA
Center on Global Energy Policy

Iran's oil buyers to reduce purchases and diversify their sources of supply. Building support to sanction other oil-producing countries in the future can be made more challenging by a US refusal to supply the global market with our own oil.

In an extreme scenario, such as global military conflict that results in widespread physical scarcity of oil, the US would always have the ability to halt crude oil exports if it is in the country's national interest to do so. Preserving crude oil export restrictions purely as a hedge against such a low-probability event is high-cost insurance.

ENVIRONMENTAL CONSIDERATIONS

While an increase in US crude oil production resulting from a modification or removal of current export restrictions has economic, security, and foreign policy benefits, it also raises important environmental concerns.

One environmental concern has been about the **local impact of increased shale development**. Indeed, development of oil and gas from shale and other tight formations poses environmental risks that must be managed at both the state and federal level. Regardless of whether the US freely exports oil, however, US oil and gas production is poised to grow sharply in the years to come, and so it is critical that states and the federal government continue to improve the level of regulation and enforcement independent of any export policy changes.

The other concern is that increased oil exports will increase greenhouse gas emissions. It is true that **to the extent oil exports boost US oil production and thus lower global oil prices, oil demand and associated carbon emissions will also rise**. Trade barriers are not an effective or appropriate response to the very real and important concerns about climate change, however.

I want to be clear: I support robust government action to address climate change. It is critically important that all nations be moving more aggressively to combat the potentially severe consequences of climate change. But **concerns about climate change are best addressed with policies targeted at that problem**. Given the economic and security benefits, **restricting oil trade is a very costly way to achieve modest greenhouse gas benefits** relative to alternatives like pricing carbon or even the EPA's power plant rules, fuel economy standards, or reducing methane emissions. It is critical that more aggressive policy actions in other areas be taken to demonstrate that boosting domestic supply, for example by allowing exports, can be consistent with meeting our climate objectives. Moreover, these other measures would deliver emissions reductions at home, while the increased emissions from allowing US oil exports would largely come outside the US.

Many government actions may raise carbon emissions, but they must be judged by weighing those costs against their benefits. For example, steps to raise GDP growth would increase energy use and emissions. Similarly, achieving a deal that prevented Iran from acquiring a nuclear weapon in exchange for allowing it to resume oil sales would likely lower oil prices and increase associated oil use and emissions. From a cost-benefit standpoint, both actions would still be desirable outcomes, notwithstanding their climate impacts. Restricting oil exports is not a cost-effective way to reduce greenhouse gas emissions.

COLUMBIA | SIPA
Center on Global Energy Policy

CONCLUSION

Today's oil market looks very different than it did in the 1970s when current crude oil export restrictions were first put in place. At that time, the United States had adopted domestic price controls to combat inflation, and crude export restrictions were necessary to make those price controls effective. While price controls have long since fallen away, crude export restrictions remain. While the magnitude and timing of the impact of easing the export restriction is uncertain, particularly given the recent oil price collapse, the direction is clear: allowing US oil exports will boost US oil supply and economic activity, along with resilience to supply disruptions, credibility in the trade realm, and geopolitical influence. While trade restrictions are not an appropriate or cost-effective way to reduce greenhouse gas emissions, it is critical that more aggressive policy actions be taken to address climate change. **The current statutory restrictions on oil exports are a legacy of a bygone era that doesn't reflect today's energy reality. On economic, security and geopolitical grounds, they should be lifted.**

Mr. POE. Thank you, Mr. Bordoff. We now turn to Mr. Kretzmann.

STATEMENT OF MR. STEPHEN KRETZMANN, FOUNDER AND EXECUTIVE DIRECTOR, OIL CHANGE INTERNATIONAL

Mr. KRETZMANN. Chairman Poe, Ranking Member Keating and members of the subcommittee, thank you very much for the opportunity to testify today. These comments are a summary of my written statement which you all should have for the record.

Oil Change International believes the crude oil export ban should not be lifted and that maintaining the ban would be helpful from the perspectives of community safety and climate protection. Our analysis predicts that lifting the ban will lead to a hazardous increase in U.S. oil production. This production would in turn likely lead to greater greenhouse gas emissions and threats to public safety such as increased crude by rail traffic.

The crude oil export ban was certainly not designed to play a role in climate change mitigation or to reduce the likelihood of a mile-long freight train full of crude oil destroying a community in America's heartland, however, it plays an important role in regulating an industry that currently has few limits placed upon it. More broadly, this issue points to the urgent need to harmonize energy policy with climate policy. We cannot drill our way out of the climate crisis, and arguments to that effect are nothing short of climate denial.

Oil Change International conducted an analysis of the impact of lifting the crude oil export ban on U.S. oil production. We estimated a projected production increase of more than 476,000 barrels per day by 2020, which incidentally is very similar to the estimate that was arrived at by the American Petroleum Institute of 500,000 barrels per day.

The critical question to consider is what will oil producers do when confronted by this additional U.S. supply? The conventional wisdom had been that OPEC would counter new supply by reducing production to support higher oil prices. This conventional wisdom has been proven completely wrong over the last year. In the past 9 months it has become increasingly clear that Saudi Arabia is determined to maintain market share rather than cut production to support higher prices. This makes the conclusion that increased U.S. production will lead to increased global production and increased emissions clearer than ever.

Lifting the crude oil export ban will likely increase crude by rail traffic putting 25 million Americans at greater risk of disaster. Since 2005, the amount of tank cars on U.S. railways has increased over 4,000 percent. At any given time there are about 135 100-car trains carrying a total of 9 million barrels of crude oil through American communities. If all of the projected increase in U.S. production were to go by rail, crude by rail traffic would see a 50-percent increase. If increased production were to reach the top end of the CGEP analysis, some 1.2 barrels of oil per day, this could more than double crude by rail traffic from today's levels.

Dozens of terminals on the Gulf Coast, at least four on the East Coast and at least six planned terminals on the West Coast, have facilities or will be designed with facilities for unloading crude oil from trains and loading it onto tankers for export. Lifting the crude oil export ban would put hundreds of communities and the lives of 25 million Americans at increased risk of an oil train disaster such as the one in Lac-Megantic, Canada, last year where 47 people perished because an oil train derailed and exploded.

It seems only a matter of luck that the incidents to date have not caused further loss of life. Crude oil trains pass through more than 400 counties including major metropolitan areas such as Philadelphia, Seattle, Chicago, Newark, Richmond and dozens of other cities. This is an already untenable situation that we cannot afford to exacerbate by creating further traffic for exports.

Lifting the crude oil export ban would also hinder progress toward the goal of climate protection. The stark reality laid out by the latest climate science is that more than three-quarters of existing proven fossil fuel reserves need to stay in the ground if the world is to maintain a two in three chance of limiting global warming to two degrees Celsius.

While it is not clear how much of U.S. oil reserves in particular need to be left in the ground, it is clear that lifting the ban would increase the incentives for production which is precisely the wrong signal to be sending. In fact, a gradual slowdown in U.S. and global oil production over time is exactly what we need in order to avoid catastrophic climate change. Any policy that could result in a net increase in global greenhouse gas emissions needs to be evaluated in terms of its climate impact.

As President Obama noted in June 2013 in regards to the Keystone XL pipeline, our national interest will be served only if this project does not significantly exacerbate the problem of carbon pollution. This climate test should be applied to all policy decisions as well as the permitting of infrastructure to extract, transport and process fossil fuels. The lifting of the crude oil export ban almost certainly fails this test. Our communities and climate in short are worth more than so-called free trade and the profits of the oil industry. Thank you very much.

[The prepared statement of Mr. Kretzmann follows:]

Stephen M. Kretzmann
Executive Director
Oil Change International

Testimony on "The Crude Oil Export Ban: Helpful or Hurtful"
Subcommittee on Terrorism, Nonproliferation, and Trade
of the House Committee on Foreign Affairs
2172 Rayburn House Office Building
April 14, 2015

Chairman Poe, Ranking Member Keating, and members of the Subcommittee, thank you very much for the opportunity to testify today.

I am Stephen M. Kretzmann, the Founder and Executive Director of Oil Change International, a non-profit charitable organization supported by over 100,000 individuals and dedicated to conducting ongoing public education regarding the environmental, social, and economic impacts associated with the production and consumption of fossil fuels.

Oil Change International believes the crude oil export ban should not be lifted and that maintaining the ban would be 'helpful' from the perspectives of community safety and climate protection. Our analysis and that of others predicts that lifting the ban will lead to a hazardous increase in U.S. oil production. This production would in turn likely lead to greater greenhouse gas emissions and threats to public safety.

Lifting the crude oil export ban would also likely lead to an increase in crude-by-rail traffic, putting hundreds of communities and the lives of 25 million Americans at increased risk of an oil train disaster, such as the one in Lac Megantic, Canada last year where 47 people perished because an oil train derailed and exploded.

The crude oil export ban was certainly not designed to play a role in climate change mitigation or to reduce the likelihood of a mile long freight train full of crude oil destroying a community in America's heartland. However it plays an important role in regulating an industry that currently has few limits placed upon it. Lifting the ban without the implementation of urgently required actions to protect the climate and communities in the path of crude oil trains or otherwise endangered by this hazardous industry, would only exacerbate these serious risks.

More broadly, this issue points to the urgent need to harmonize energy policy with climate policy. We cannot drill our way out of the climate crisis, and arguments to that effect are nothing short of climate denial.

Lifting the Crude Oil Export Ban will increase U.S. oil production

The key reason that U.S. oil producers want an end to the export ban is to gain access to international markets, thus raising the price they receive for their crude. In recent years, a glut of U.S. light crude has caused a structural price differential between North American crude oil and international crude oil. This is primarily manifested in the spread between the crude oil benchmarks: WTI and Brent. Exporting U.S. crude oil would essentially end the glut of U.S. crude within in the North American market and raise the price of WTI, even while the entry of U.S. crude into the international market may lower the price of Brent.

Raising the price producers receive for their crude facilitates greater production by raising capital available to reinvest in new production and by bringing into play oil fields that may not have been economic with lower crude oil prices. Allowing exports will also simply create a larger market for U.S. crude than would otherwise be available. The end result of all these factors is a hazardous increase in U.S. oil production.

In March 2014, Oil Change International conducted an analysis using Rystad Energy's UCube database to estimate the potential production increase caused by a $10 per barrel increase in the price received by U.S. oil producers.[1] We found that a $10 per barrel increase in U.S crude oil prices could stimulate an additional 9.9 billion barrels of crude to be produced between 2015 and 2050.

At almost the same time, the American Petroleum Institute (API) estimated that U.S. production could increase by 500,000 barrels per day (bpd) by 2020 if the export ban is lifted.[2]

Oil Change International's analysis estimated an average projected U.S. oil production increase of more than 476,000 bpd by 2020, which is very similar to the API estimate of 500,000 bpd by that time.

A more recent report by the Center on Global Energy Policy (CGEP)[3] gives a wider range but certainly demonstrates that lifting the U.S. crude oil export ban could indeed increase U.S. production and place additional crude oil supply onto the world market.

The critical question to consider is: What will other oil producers do when confronted by this additional U.S. supply? The conventional wisdom had been that the Organization of the Petroleum Exporting Countries (OPEC) would counter new supply by reducing production to support higher oil prices. This conventional wisdom has been proven spectacularly wrong over

---

[1] Oil Change International. "Lifting the Bank, Cooking the Climate: The Climate Impact of Ending the U.S. Crude Oil Export Ban." March 2014. http://priceofoil.org/content/uploads/2014/03/LiftingTheBanFinal.pdf Although this was before the oil price crash that ensued later in the year it was only designed to be indicative of the impact of lifting the export ban on production and thus remains valid today.

[2] ICF International and EnSys Energy. "The Impacts of U.S. Crude Oil Exports on Domestic Crude Production, GDP, Employment, Trade, and Consumer Costs The Impacts of U.S. Crude Oil Exports on Domestic Crude Production, GDP, Employment, Trade, and Consumer Costs." American Petroleum Institute, March 31, 2014. http://www.api.org/rss/~/media/Files/Policy/LNG-Exports/LNG-primer/API-Crude-Exports-Study-by-ICF-3-31-2014.pdf

[3] Jason Bordoff and Trevor Houser. "Navigating the U.S. Oil Export Debate." Columbia University School for International and Public Affairs Center on Global Energy Policy, January 2015. http://energypolicy.columbia.edu/sites/default/files/energy/Navigating%20the%20US%20Oil%20Export%20Debate_January%202015.pdf

the last year. In the past nine months it has become increasingly clear that OPEC's most important member, Saudi Arabia, is determined to maintain market share rather than cut production to support high prices. This makes the conclusion that increased U.S. production will lead to increased global production clearer than ever.

Lifting the Crude Oil Export Ban will likely increase crude-by-rail traffic - putting 25 million Americans at greater risk of disaster

The U.S. oil boom has precipitated a parallel boom in the transportation of crude by rail. Since 2005, the amount of tank cars on the U.S. railways has increased over 4000%.[4] Around 1 million barrels of crude oil is currently loaded and unloaded onto and off of the rail network every day in the United States.[5] With the number of days from source to destination averaging 9 days, this means that at any given time there are about 135 one-hundred-car trains carrying a total of 9 million barrels (378 million gallons) of crude oil through American communities at any given time.[6]

However, the capacity of loading and unloading terminals in the U.S. and Canada is nearly five-times greater.[7] In addition, planned new terminals and capacity expansions at existing terminals may add a further 1 million bpd in the next two years.[8]

Lifting the crude oil export ban would raise U.S. production further and likely send more crude oil trains to terminals on the East, West, and Gulf Coasts for export. If all of the projected increase in U.S. production (500,000 bpd) were to go by rail to export terminals crude-by-rail traffic would see a 50% increase. If increased production were to reach the top end of the CGEP analysis – some 1.2 million bpd – this could more than double crude-by-rail traffic from today's levels.

Dozens of terminals on the Gulf Coast, at least four on the East Coast and at least six planned terminals on the West Coast have facilities, or will be designed with facilities, for unloading crude oil from trains and loading it onto tankers for export. This already occurs in Albany, New York, where trains are unloaded to barges that send crude oil down the Hudson River for export to Canada. This has generated concern in Albany and right through the Hudson Valley resulting in a moratorium on the expansion of crude-by-rail in Albany.[9]

American citizens are rightly concerned about current crude-by-rail activity and even more concerned about the potential for it to grow further. According to a review of federal accident records conducted by Associated Press, at least 21 oil-train accidents and 33 ethanol train accidents involving a fire, derailment, or significant amount of fuel spilled have occurred in the

---

[4] Association of American Railroads. "Crude Oil By Rail." Accessed on April 13, 2015. https://www.aar.org/todays-railroads/what-we-haul/crude-oil-by-rail

[5] U.S. Energy Information Administration. "U.S. Movements of Crude Oil By Rail." March 30, 2015. http://www.eia.gov/petroleum/transportation/

[6] Oil Change International. "Runaway Train: The Reckless Expansion of Crude-by-Rail in North America." May 2014. http://priceofoil.org/content/uploads/2014/05/OCI_Runaway_Train_Single_reduce.pdf

[7] Oil Change International. "Runaway Train: The Reckless Expansion of Crude-by-Rail in North America." May 2014. http://priceofoil.org/content/uploads/2014/05/OCI_Runaway_Train_Single_reduce.pdf

[8] Oil Change International. "Runaway Train: The Reckless Expansion of Crude-by-Rail in North America." May 2014. http://priceofoil.org/content/uploads/2014/05/OCI_Runaway_Train_Single_reduce.pdf

[9] Earthjustice. Albany County Halts Expansion of Dangerous Crude by Rail Project. March 12, 2014. http://earthjustice.org/news/press/2014/albany-county-halts-expansion-of-dangerous-crude-by-rail-project

U.S. and Canada since 2006.[10] This does not include the five incidents that recently occurred in February and March 2015.[11]

In July 2013, 47 people were killed in the small town of Lac Megantic, Quebec when a train carrying crude oil from North Dakota derailed and exploded in the middle of the town. Since then, at least ten major incidents have occurred in the U.S. and Canada involving derailed crude oil tank cars and serious explosions and fires.

Over 25 million Americans live within the 'blast zone' of crude oil trains. This is an area of within 1 mile from the tracks.[12] It seems only a matter of luck that the incidents to date have not caused further loss of life.

Crude oil trains pass through more than 400 counties, including major metropolitan areas such as Philadelphia, Seattle, Chicago, Newark, Richmond, and dozens of other cities.[13]

A recent Department of Transportation report estimated that an average of ten derailments will occur annually for the next two decades.[14] This is an already untenable situation that we cannot afford to exacerbate by stimulating further traffic for exports.

The Crude Oil Export Ban is not a climate policy, but lifting it would hinder, not help, progress toward the goal of climate protection

In the current market, additional U.S. oil production will likely lead to an increase in greenhouse gas emissions. Every additional barrel of oil produced and consumed emits from 550 kg to 850 kg of carbon dioxide equivalent, depending on the type of oil that is being produced and consumed.[15]

There are two primary ways in which an increase in U.S. oil production harms the climate:

- Failure to keep oil in the ground,
- Increased demand brought about by greater supply.

The stark reality laid out in the Intergovernmental Panel on Climate Change's (IPCC) Fifth Assessment report of November 2014 is that more than three quarters of existing, proven fossil

[10] Matthew Brown and Josh Funk. "Fuel Trains Could Derail Up To 10 Times A Year Over Next Two Decades, Feds Predict." Huffington Post, February 22, 2015. http://www.huffingtonpost.com/2015/02/23/ap-exclusive-fuel-haulin_n_6730476.html

[11] Matthew Maiorana, "How many explosions before we stop crude-by-rail?" Oil Change International Price of Oil Blog, March 13, 2015. http://priceofoil.org/2015/03/13/many-explosions-will-take-stop-crude-rail/

[12] Forest Ethics, "Oil Train Blast Zone Website." Accessed on April 13, 2015. http://explosive-crude-by-rail.org/

[13] Matthew Brown and Josh Funk. "Fuel Trains Could Derail Up To 10 Times A Year Over Next Two Decades, Feds Predict." Huffington Post, February 22, 2015. http://www.huffingtonpost.com/2015/02/23/ap-exclusive-fuel-haulin_n_6730476.html

[14] Matthew Brown and Josh Funk. "Fuel Trains Could Derail Up To 10 Times A Year Over Next Two Decades, Feds Predict." Huffington Post, February 22, 2015. http://www.huffingtonpost.com/2015/02/23/ap-exclusive-fuel-haulin_n_6730476.html

[15] Deborah Gordon, Adam Brant, Joule Bergerson, and Jonathan Koomey. "Know Your Oil: Creating a Global Oil-Climate Index." Carnegie Endowment for Peace, March 11, 2015. http://carnegieendowment.org/2015/03/11/know-your-oil-creating-global-oil-climate-index/i3v1

fuel reserves need to stay in the ground if the world is to maintain a 2 in 3 chance of limiting global warming to two degrees Celsius.[16]

While it is not clear how much of U.S. oil reserves in particular need to be left in the ground, it is clear that lifting the ban would increase the incentives for production, which is precisely the wrong signal to be sending. In fact, a gradual slow-down in U.S. – and global - oil production over time is exactly what we need in order to avoid catastrophic climate change. The International Energy Agency has also warned that "no more than one-third of proven reserves of fossil fuels can be consumed prior to 2050 if the world is to achieve the 2°C goal,"[17] which is the conservative, globally accepted threshold of average global temperature increase for avoiding catastrophic climate change.

As fossil fuel production is increasing in the U.S. and globally, our window to meet this target is closing fast (See Figure 1).

Figure 1. The carbon content of fossil fuel reserves in comparison to the carbon budget[18]

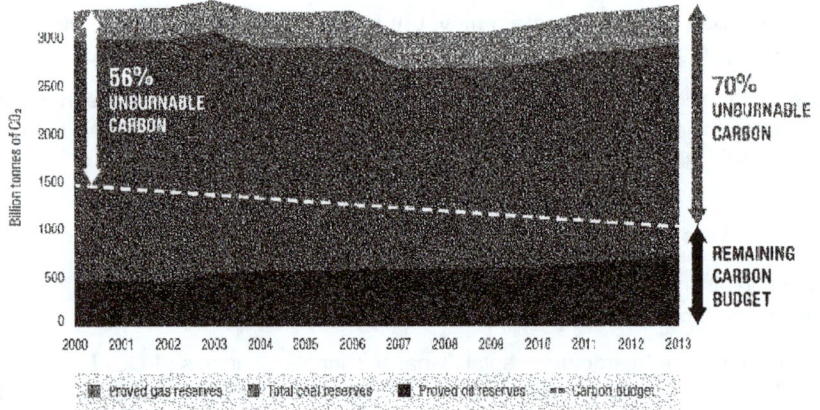

The percentage of total fossil fuel reserves that are unburnable has grown rapidly over the past decade: proven global oil, gas and coal reserves have risen while the carbon budget (the amount left to burn) has shrunk as the result of rising greenhouse-gas (GHG) emissions. Source: All data from U.S. EIA, IPCC, and Global Carbon Project - calculations by Oil Change International.

In the same Fifth Assessment report, the IPCC notes that not only will it be necessary to keep carbon in the ground, but that in fact global emissions must fall dramatically between now and

---

[16] Intergovernmental Panel on Climate Change. "Climate Change 2014: Synthesis Report. Contribution of Working Groups I, II and III to the Fifth Assessment Report of the Intergovernmental Panel on Climate Change." Geneva, Switzerland, 151 pp. https://www.ipcc.ch/report/ar5/syr/

[17] International Energy Agency (IEA). "World Energy Outlook 2012." http://www.iea.org/publications/freepublications/publication/English.pdf

[18] Source: Elizabeth Bast, Shakuntala Makhijani, Sam Pickard and Shelagh Whitley, "The fossil fuel bailout: G20 subsidies for oil, gas and coal exploration." Oil Change International and Overseas Development Institute, November 2014.

2050. Lifting the export ban, which is likely to increase global oil demand, clearly pushes this in the wrong direction.

How much additional global oil demand may be stimulated by liberalized U.S. oil exports is subject to many factors. The CGEP report suggests a range of between 0 and 1 million bpd – a wide range derived from the multiple uncertainties considered in the analysis. However, if we assume that a likely impact that is in the middle of that range, then an additional 500,000 bpd of additional oil demand would lead to emissions of up to 110 million metric tons of carbon dioxide equivalent (CO2e) per year. This is equivalent to the emissions from 29 average US coal-fired power plants or over 23 million average passenger vehicles.[19]

A net increase in global greenhouse emissions is very likely to be the result of lifting the crude oil export ban. The United States and the world have agreed in multiple international forums to limit average global temperature rise to below 2 degrees Celsius. At this point, the world is dangerously close to passing the point at which that goal can be achieved and therefore condemning future generations to climatic changes that will drastically challenge their chances of living prosperous and secure lives.

Given this context, any policy change that could result in a net increase in global greenhouse emissions needs to be evaluated in terms of its climate impact. As President Barack Obama noted in June of 2013 in regards to the Keystone XL pipeline:

> [O]ur national interest will be served only if this project does not significantly exacerbate the problem of carbon pollution."[20]

This 'climate test' should be applied to all policy decisions as well as the permitting of infrastructure to extract, transport, or process fossil fuels. The lifting of the crude oil export ban almost certainly fails this test.

Conclusion

The crude oil export ban was not designed to mitigate climate change or to reduce the likelihood of a freight train full of crude oil destroying a community in America's heartland. However, in the absence of adequate regulation to mitigate either of these crucial issues it plays an important role in regulating an industry that has few limits placed upon it. Lifting the ban prior to implementing the urgently required action to protect the climate and protect communities in the path of crude oil trains, can only exacerbate these serious risks.

---

[19] EPA Greenhouse Gas Calculator: http://www.epa.gov/cleanenergy/energy-resources/calculator.html

[20] The White House Office of the Press Secretary. "Remarks by the President on Climate Change." Georgetown University, Washington, D.C., June 25, 2013. https://www.whitehouse.gov/the-press-office/2013/06/25/remarks-president-climate-change

Mr. POE. Thank all our panelists. I will begin with myself asking questions. Try to limit questions by the members to 5 minutes, and so therefore you have 5 minutes to answer these questions.

Start with you, Ms. Rosenberg. We have lifted, so to speak, the sanctions, the ban on Iran for exporting crude oil. That would be the long term policy if this deal goes through allowing Iran to export some of their crude oil. Does it make sense to you that we would allow Iran to put more oil on the world market but still prohibit America from putting more oil on the world market?

Ms. ROSENBERG. Thank you for the question, Mr. Chair. In fact, I don't think that makes sense at all. The U.S. has the greatest degree of leverage and influence in the market if it allows its producers to produce and sell their oil in an open, international market. Then the United States will be in a better position to, if necessary, if the additional sanctions or the reimposition of sanctions is appropriate in a policy circumstance, be able to quickly move, to credibly impose that policy and ask international allies to join with the United States in doing so which of course represents a sacrifice on their own part. They will be looking for alternative supplies to enter the market in order to go along with that policy. Lifting the crude exports ban will help make that a reality for them.

Mr. POE. Let me talk about our allies. Countries need obviously crude oil imports. Europe is a primary example. And they have mentioned to me that it seems to be we want them to support sanctions against a country so they can't export, but we don't provide them an alternative for importing crude oil from the United States. That seems to be our policy. Would that help our ability to deal with our allies in an easier way if they had an alternative for, okay, you want us to have sanctions on Iran where we get oil, but you don't provide us crude oil. Do you think that would be a better policy to say, okay, here comes the Cavalry? We are going to supply you some Texas crude oil.

Ms. ROSENBERG. I think such a policy would put the United States in a much stronger position to encourage and influence our allies to join with the United States in imposition of sanctions. And experience from the Iran case would certainly bear that out where international allies said to the United States, this is very difficult for us economically. We join in this policy because we think it is the right foreign policy measure, but we have come to a point, or are near a point, where we can go no further unless there are alternative supplies.

Mr. POE. Russians, well, the Europeans get about 40 percent of their energy oil from Russia. How would lifting the export ban on America thwart Russian monopoly, aggression maybe, policy, how would that impact it in your opinion?

Ms. ROSENBERG. Lifting the——

Mr. POE. U.S. ban.

Ms. ROSENBERG. Lifting restrictions on U.S. exports.

Mr. POE. U.S. ban on American exports, not on Iranian exports.

Ms. ROSENBERG. Exactly. What it will do is stimulate greater supply from the United States, and greater diversity in the international supply pool, which will make Russia work much harder to supply Europe with oil. That will reduce its revenue if it plans to keep its market share, which is certainly in line with U.S. policy toward Russia which involves degrading their revenue generation ability in the energy sector in the medium and long term.

Mr. POE. The United States is critical of China for not exporting its rare earth minerals. There may even be a complaint with the World, WTO, I am not sure. Are we somewhat hypocritical by criticizing China for not exporting but yet we don't export our energy?

Ms. ROSENBERG. Not only hypocritical but also poorly placed to influence other countries to embrace free trade policies not just on energy but on other natural resource commodities. More broadly, at a time when the United States is engaged in very serious, significant discussion about free trade arrangements with Atlantic and Pacific partners, now is the opportunity for the United States to be sending the signals to some of those countries particularly in East Asia that will be making trade related decisions that will in fact impact our economy in the decades to come.

Mr. POE. Thank you.

Mr. Kretzmann, you mentioned that the U.S. policy, our U.S. policy, of not exporting crude oil is because of the environment concerning climate change or to protect the environment. More exports would allow the environment not to be as good as it should be. I assume then that your organization then recommended to the administration that they not lift the ban on Iran because Iran now will be able to produce more energy and therefore they will pollute more. And so did you make this recommendation to the administration that they don't lift the ban on Iran because they are going to produce more and it is going to hurt the climate?

Mr. KRETZMANN. In general, we actually believe oil production across the globe should be phased down to levels that are consistent with the climate challenge. And so, no, we didn't make a specific recommendation on Iran, but we weren't called on to either. And we are quite consistent——

Mr. POE. Maybe that is why they didn't.

Mr. KRETZMANN. We are quite consistent—yes, I am sure. We are quite consistent on the fact that less oil needs to be used and we need to keep oil production and consumption to limits that are prescribed by global climate science.

Mr. POE. The other comment that you made is that more rail cars are in America now because there is more oil. We could probably diminish the rail car capacity if we had more pipelines. That is just an observation. But my time is expired. I will yield to the ranking member.

Mr. KEATING. Thank you, Mr. Chairman. It is great to know that in the whole scope of things our sanctions with Iran deal more with, much more with their ability to have a nuclear weapon than international oil prices.

We talked about the effect and I think it really can't be questioned that lifting the ban would create more jobs. But in my state of Massachusetts, and I know how Texas has such an interest in this and I understand that too, in my own state one of the fastest growing industries has been surrounding the renewable energy. And I would just like to ask what effect would this have on the growth of renewable energy, wind energy, solar energy, geothermal in our country? I will ask Mr. Kretzmann.

52

Mr. KRETZMANN. I think unfortunately ongoing policies that support the further growth of the oil industry in the United States tend to at least diminish investment in renewables, although on the other hand we are seeing more than we have ever seen before and that is good news. But I think the sooner that we can make it clear that our energy future is about renewables, and oil and coal and natural gas are about our energy past, the faster we can move markets to create more investment to create that transition that we all know is coming.

Mr. GRUMET. Mr. Keating, Massachusetts did a profound job of advancing renewables but the vast majority of it is in the electric power section. The wind power is displacing coal and natural gas. We have almost no oil left in the electric sector. So I think the renewables question here really relates not so much to wind or solar or bio, but to biofuels and ethanol. And I think there is a conversation to be had about how domestic oil prices create or discourage market for alternative transportation fuels, but really not so much of a conversation about the interaction with wind power.

Mr. KEATING. Because I have noticed in Europe how they are ahead of us in so many of these other areas as well and I assume some of their problems with access to oil and their concerns that were referenced by our witnesses today also spurred growth there in renewables.

But I would just too like to touch base on the effect economically on our domestic refineries. What effect would this have on our refineries and could you put it in the context of what has been happening over the last few decades with our domestic refineries too? I will address that to anyone that wants to take that.

Mr. GRUMET. I will take a shot, and Jason Bordoff knows more about it than I do so he will go next. I think the most important point I would like to make is that honestly none of us know. And it is critically important to recognize that the question here is not should we mandate oil exports, the question is should we step back and actually let that truth reveal itself?

There are arguments made that in fact the most economic outcome would be for the refining industry to make significant investments here at home so it could process all of our domestic oil. The depressed price right now suggests that is not the case at the moment, but it is entirely possible that could be true. It is not my guess, but it could be true. The only way we figure that out is if you lift the restriction and let the truth become the truth.

I think the mistake here, and we have often found there is a certain seduction to wanting to know the answer which encourages us all to want to try to pick market outcomes, the history of Congress picking market outcomes has not actually been a very proud one. And so I think most broadly, the best thing that we could do is not try to figure out whether it is the producers or the refiners who have the best of this projection and, in fact, let the market make that determination.

Mr. KEATING. Yes, and if the refiners have that discount. And I would just like to ask you, what has been happening with domestic refineries? Have they contracted? Have they become more scarce? And if that is the case what effect could this have on that domestic job industry?

Mr. BORDOFF. I can take a crack at answering it. I mean I think we know a little bit what the impact would be. As we heard, the U.S. refineries have been running at a very high utilization rate. The U.S. has gone from being the largest importer of refined petroleum in the world in 2006 to the largest exporter today. That is a dramatic turnaround.

It is true that we saw refineries that depended on light crude, particularly in the Northeast, at risk of closing several years ago, and some were kept open because they were acquired by airlines or private equity firms or others and they have been enjoying a benefit of the discount that we are talking about, so this is sort of the direct corollary of the issue we are talking about.

To the extent that this is as big of an issue for production as many producers say, it is because there is a discount that would otherwise develop in the market and that would benefit some refineries. If you don't allow that discount to develop they won't enjoy access to discounted crude and they won't have that benefit. The question is whether, if you are concerned about the access to refined petroleum product in the Northeast, you think the right policy approach is to put in place an economy-wide restriction on energy trade in order to create a price discount that is, in effect, a subsidy to some refineries that are economically challenged, or whether it makes sense to allow energy trade and develop other policies to promote security of energy supply for refined petroleum in the Northeast. The administration recently put in place a gasoline product reserve in the Northeast. There are a host of other measures that one could put in place as well.

Mr. KEATING. Thank you. I yield back, Mr. Chair.

Mr. POE. I thank the ranking member. The chair will now recognize the gentleman from Wisconsin, Mr. Ribble.

Mr. RIBBLE. Good morning, everybody, and thank you for being here. Mr. Bordoff, have you done any research in your study about the comparative regulatory regimes in environments around the globe as it relates to oil production? In other words, is the U.S. regulatory regime regarding cleanliness, safety and what not, similar to that that is going on in Venezuela, Iran, Russia, Saudi Arabia? Could you talk about that for a minute?

Mr. BORDOFF. We didn't look at it in this study. I would say the answer is generally no. I think the U.S. has quite high regulatory and safety and environmental standards relative to many of the countries you just mentioned for oil production.

Mr. RIBBLE. Yes, those are the primary other competing nations, correct? Did I miss anybody there? Canada, I suppose.

Mr. BORDOFF. Yes, and Canada. If we think about where U.S. refineries are getting their oil from and the heavy crude that they are optimized to run we would be on the margin importing a little more from Canada and Venezuela along with Mexico and some others. Yes.

Mr. RIBBLE. Ms. Rosenberg, do you have any comments on that?

Ms. ROSENBERG. Nothing further, thank you.

Mr. RIBBLE. Okay, thank you.

Mr. Kretzmann, in light of that would you agree with Mr. Bordoff's assessment?

Mr. KRETZMANN. I think that there is certainly great room for improvement in U.S. oil production standards.

Mr. RIBBLE. That wasn't my question.

Mr. KRETZMANN. But I hear your question and I think U.S. oil production standards are often higher than in some other countries, although perhaps not absolutely the top of the class worldwide. That said, I think the question implies——

Mr. RIBBLE. Who is top of the class worldwide?

Mr. KRETZMANN. Norway, I think, is pretty good, and actually Brazilian offshore is also really good. But besides that the question implies that somehow U.S. oil production would replace other oil production, and the experience in the market over the last 6 months makes it clear that it will just be added to ongoing production. And so——

Mr. RIBBLE. The question didn't imply that at all. The question implied that U.S. regulatory regime is more stringent than other countries. That is all the question implied.

Mr. KRETZMANN. Oh, I certainly assumed that what you were saying was wouldn't you rather have U.S. oil than other oil, and that seems to be a reasonable conclusion to pull. And my point is, it is not a question of one or the other, it is actually going to be both.

Mr. RIBBLE. But in light of that though then wouldn't you say you would rather have cleaner production than less clean production?

Mr. KRETZMANN. I would rather have less production.

Mr. RIBBLE. I get that. You would rather have no production.

Mr. KRETZMANN. Ultimately I would rather have production that brought us within climate limits.

Mr. RIBBLE. Right. And I don't know that in light of that, the approach of saying, ''Let us get rid of production'' is the fastest way to get at a cleaner climate. I think the faster way is to get at the user than it is to the producer. Producers respond to demand as opposed to creating demand. They respond to it. And if you can change that paradigm you are likely to reduce demand. However, I would say until that time comes—and let us face it. With CAFE standards and other things that have happened, things have gotten better rather than worse in light of carbon use.

It just seems to me it would behoove global climate interest to have production happening in places that are cleaner and safer as opposed to places that are dirtier and less safe. And so it is almost like you are arguing against your position here.

Mr. KRETZMANN. No, not at all. Producers actually impact demand quite a bit as we have seen over the last year. I mean the increase in production that has happened since the U.S. increase in production, which has been quite substantial over the last several years, and then the Saudis not responding by reducing their production has actually significantly lowered the price as we see and that in turn has increased global demand. And so producers obviously impact global demand, and that is an important part of the equation for us to consider from an economics perspective when thinking about how to influence the market. I mean this is a complex challenge about how to begin to wind down our global addiction to fossil fuels in order to respond to climate change. I think we all recognize that. But there are things that we have to do both on the supply and demand side in order to meet that challenge.

Mr. RIBBLE. Yes, and I guess my only challenge to you in your thinking, and it is not necessarily related to the trade issue, is that as you look in the long view it seems to me that you are better off really accepting the fact that between now and where you would like to be there are maybe better processes to get there than the one that you are pursuing at this particular moment. That would just be my suggestion.

Mr. Grumet, could you also help me understand a little bit better on how you come to the conclusion that gas prices would be lower even though oil prices would be higher?

Mr. GRUMET. I don't believe my testimony indicates that global oil prices would be higher.

Mr. RIBBLE. But the U.S. produced oil would be sold at a higher price.

Mr. GRUMET. The U.S. produced oil would be sold at the global market price. And I think the question for gasoline prices is the relationship between the refinery and the ultimate market. So the assumption that a discounted crude price, a refinery in the U.S. can get crude at $6 or $7 a barrel less than someplace else, is that that will somehow again altruistically be passed on to you and me. And that would be an irresponsible choice by a refiner with obligations to shareholders.

What a refiner should do is seek the best price for their product. Because they of course appropriately have access to a global market, they will get the same price as any of the global competitors. And so the challenge of course is to find a way to have the benefits of a robust global market that creates consumer benefits and foreign policy benefits, and at the same time make sure that we have a dynamic economic situation here at home.

Mr. RIBBLE. All right, thank you. I yield back.

Mr. POE. I thank the gentleman. And the chair recognizes the gentleman from California. Mr. Sherman?

Mr. SHERMAN. I have a lot of comments here. It may eat into my question time so the witnesses can relax for awhile. I talk to a lot of Ambassadors, as we all do, I never bring up oil. Oil is not—in sense of oil production, but I just ask them what is on their mind and none of them have ever said, gee, the most important thing is U.S. exports of oil. So I am not sure that we are going to get huge concessions on other trade issues by adopting the policies that three of our witnesses would like.

As to the environment, environmentalists have not focused on the tremendous harm to the environment of wars in the Middle East. And the fact is that lower energy prices worldwide will drive down the power of the those who cause those wars. The most extreme example, well outside our discussion here, was already brought up, and that was Iran. Yes, Iran may be producing another million barrels a day. That may be bad for the environment. But if we reduce by 1 percent the likelihood that nuclear weapons are used and you weigh that against 1 million barrels a day, now I am not sure which you go to reduce the chance of a nuclear Iran, but whatever we can do to do that is a plus for the environment even if it means 1 million barrels a day.

The idea that we should have bottlenecks and problems and re-strictions to drive up the price of oil so that we have less carbon, maybe we should get rid of the bottlenecks and the problems and then have a tax to support infrastructure and that way we can have infrastructure instead of bottlenecks.

As to rail transport, the only thing worse than bringing a train through my district loaded with crude is to bring a train through my district loaded with refined gasoline. And so if we encourage moving a lot of gasoline around so we can export it that might be worse. But I would hope that as part of the bill we put together we go even further than the administration already has in terms of safety of our oil by rail transportation. I think we are being given kind of a false choice. It is, stick with the present policy, which is kind of crazy, or lift all the barriers. Well, one thing we might do is go completely the other way and that is ban the export of refined petroleum. Make them give the discount to people who live in my district. That ought to be explored so that we can go in both directions.

We can also take a look at our rules for exporting natural gas, which I know are unpopular with some members of the committee. But perhaps we can just go to that point instead of completely lift-ing the ban, and have the administration have to license the ex-port. And I would point out that in the natural gas area, Ameri-cans are paying much less than the world price. Part of that is physics and the cost of transportation, but part of it might be our limitations on export.

For those who support this idea, you might want to make it a little easier for us to vote for it. When people hear this they hear a threat to the security and price of the oil and the gasoline they buy at the pump. So you may want to explore the idea of expand-ing swaps and making it practical. One could imagine a situation where if you bring in a barrel of crude you get a chit. And if you want to export a barrel of crude you need a chit. I think the price of these chits would be about a penny a barrel, and that solves 99 percent of your problem. Because if we can turn to people and say, yes, this is a swap, for every barrel we are exporting we are im-porting a barrel, a company is importing and exporting or is in partnership with the import or the export. That is very different than saying that you are going to take that oil from North Dakota or wherever, bring it to the Port of Los Angeles right by my thirsty consumers, and ship it to Japan, if that is the only part of the pic-ture.

We drive a lot in Southern California and that is, if this is part of a system for more efficient refinery and more efficient pricing that is swaps, that is a lot better picture to put forward than to see all that oil leave our country, until of course we all buy Teslas and then we will be able to do it. And I look forward to that day and I yield back.

Mr. POE. I thank the gentleman from California. The chair recog-nizes the gentleman from South Carolina, Mr. Wilson.

Mr. WILSON OF SOUTH CAROLINA. Thank you, Judge Poe, and thank each of you for being here today.

Mr. Grumet, do you believe that lifting the crude oil export ban would bolster the U.S. negotiating position on other trade issues?

Mr. GRUMET. Mr. Wilson, thank you. I think that is an important point that a number of us have spoken to, and again it goes back to kind of a fundamental value question. The United States of America has always been, I think, the global leader in advocating for open markets, efficient and free trade. And we have a little bit of a challenge. There is a bit of a hypocrisy in suggesting to others that they share their riches with the world but we are somehow going to hold tight on this commodity.

And again I think Mr. Bordoff makes the important point, that this is a policy that was adopted in a very different environment. It really was not designed to kind of be Fortress America, but that is the effect it is now having, and now in fact it actually does matter. Now we have a profound opportunity through this remarkable abundance that none of us predicted to reassert a voice in a global economy in a very, I think, challenging global environment.

This is benefiting us in a myriad of ways. It is benefiting us in trade. It is benefiting us in our ability to provide opportunities to our European allies to fend off some of the manipulations of the Russians. It is enabling us to hold together coalitions around sanctions. So I think you have heard a lot of consistency at least from the first three witnesses that there is a very significant advantage to being part and a forceful player in this global market.

Mr. WILSON OF SOUTH CAROLINA. And thank you for your points. And Ms. Rosenberg, I am particularly interested in lifting the crude oil ban. What would be the effect on jobs? And I put it in the context of Keystone Pipeline that was with our Canadian allies, the bringing of crude product through our country. I know personally that almost 1,000 jobs, permanent jobs, would be created in a community that I represent.

Michelin Tires makes the tires for Fort McMurray, Alberta, Canada. They are 12 feet high, $60,000 each, and nearly 500 jobs. And then MTU makes engines for the processing of the oil sands. Again you could get three engines in this room. They are very nice engines. And again nearly 500 jobs. A total of 1,000 jobs just in the district I represent. And so by lifting the ban, what would be the effect on jobs across our country?

Ms. ROSENBERG. Thank you for the question. Essentially, lifting restrictions on crude exports sends a signal to those producers who find it difficult to access international benchmark prices for their crude to be able to access them, which drives an incentive for them to expand production, expand their market share. Broadly speaking, what that does is create additional jobs for those producers and for associated industries that support them in services in those communities, et cetera.

Now the number of jobs that that will create there are different studies, they offer different numbers, and it is important to remember this is also a function of what the international oil price is. If it is particularly strong that will incentivize greater investment, and that is of course a cyclical, it is a cyclical market that moves up and down. I would defer to Jason Bordoff whose study has a particular comment on this if he wants to speak to it, but in broad terms it is true that job creation would be a function of lifting export restrictions.

58

Mr. WILSON OF SOUTH CAROLINA. And Mr. Bordoff, again an excellent report. And on the issue of jobs what do you see?

Mr. BORDOFF. Well, our study didn't estimate a particular number of jobs, but generally we had a range of the impact it could have on production. Our estimate was anywhere from zero to 1.2 million barrels a day and it depends on a host of circumstances including how quickly U.S. production grows, how much of a price discount might otherwise emerge, how quickly refineries can adapt.

But it is a function of how big an impact this has on U.S. production; so to the extent the export restriction is eased, the more significant an impact it has on increasing U.S. production, the more economic activity, the more employment you are going to see in the oil and gas sector in the U.S. The magnitude and timing of that impact is frankly highly uncertain, particularly given the price collapse that we have seen and what is happening to U.S. production, but directionally it is going to be positive.

Mr. WILSON OF SOUTH CAROLINA. Well, and I know personally, in fact my family, I understand the opportunities provided by the oil industry, the liberating, fulfilling lives that people can have. My great grandfather started with Standard Oil New Jersey in Virginia in 1895. My grandfather was the division manager at Standard Oil of New Jersey in South Carolina Esso, and then my dad was a sales representative for Exxon Humble, and I am very grateful in them. My brother was an oil jobber.

So I know that the oil industry is very important to providing opportunity and I appreciate every effort to expand it for the benefit of jobs and opportunity and fulfilling lives to the American people.

Mr. POE. Yield back?

Mr. WILSON OF SOUTH CAROLINA. Yes, I do.

Mr. POE. All right. The chair recognizes the gentleman from Texas, Mr. Castro.

Mr. CASTRO. Thank you, Chairman Poe, and thank you to each of the witnesses for your testimony this morning. Of course this is an issue. This is a policy, a longstanding policy, 30 years or so. We seem to be taking on a lot of these issues lately with Cuba being another one, and it seems to me that there are basically three issues here—one of geopolitics, another of environmental concerns, and third, the domestic business consideration that has expressed itself as a battle between producers and refiners. But I guess with those three things in mind, I have a few questions.

First of all, if we were going to put together a bill that would lift the ban as it exists, whether it is a partial lifting or a complete lifting, what would the safeguards look like, right? So, for example, what if we were in a situation like the 1970s again where you had a scarcity of resources? What would the safeguards be that we would need to put in place to make sure that we don't go through a situation like that again? I mean there was a reason that this ban was put in place back then, right? If we are in a situation like that again what do we do?

And then the second part is, since this is kind of the first round of discussions about this, if you were going to design a grander bargain, a larger bill where you would allow for some perhaps partial lifting of this ban, but also an infusion of resources or the support of policies to develop alternative energies, what would that look like? What would a grander bargain look like? Please.

Mr. GRUMET. Well, as the Bipartisan Policy Center we love grand bargains, Mr. Castro, so I am going to take a shot, I think maybe try to address the second part of your question first. I think the first point is that while the benefits economically, I believe, strongly outweigh the costs, and there will be costs, there will be a handful of refiners, a few in the Northeast, who are clearly going to have to struggle with this recalibrating market. And I believe that the Congress is going to have hopefully look to opportunities to smooth that transitional challenge. I have yet to see a specific proposal to do so.

There is not a lot of conversation: I mean I love phasing and all those kinds of ideas, but I think there is a certain reluctance upon the part of those who feel that they might be injured to suggest a path forward because they believe it will in fact make it easier for you to pursue that path. I think as the debate becomes more and more serious and this becomes, I think, what it will be which is an inevitable move to change policy, I think those ideas will come forward. I certainly hope——

Mr. CASTRO. What would you do? What safeguards would you put in place?

Mr. GRUMET. You are asking another important question which is what happens if something changes, right? I mean all of a sudden the oil market is certainly proven to be volatile. The President, in current law, has significant authority to make decisions in the national security interest to right now allow exports in contradiction to the ban, and I certainly think that that authority should be mandated or, sorry, preserved and even strengthened in reverse. So once this market, I think, is opened and we have the opportunity to engage in the international stage, the President certainly must have a sustained authority to interrupt those exports if necessary to ensure the security interests of the nation. And so I think the way one crafts that is important, but that it an important aspect of this debate that I certainly hope Congress continues to pursue.

Mr. CASTRO. Sure.

Ms. ROSENBERG. Following on that briefly, I think there are two main elements of the question that you just asked. And the first one is how do we make the U.S. economy its most resilient version in order to protect U.S. consumers from the circumstances of the 1970s that were so economically painful, particularly deriving from energy sector changes, dramatic changes?

And the second element of your question I see is that what are the reassurances that you could put, the policy measures particularly including reassurances that you could put into a piece of legislation that give consumers the confidence to know that in fact this policy is consistently on an ongoing basis performing in a way that benefits them and additionally that maintains for the executive branch the ability to put the brakes on if circumstances merit?

And speaking to that particular issue, the way to provide information, there are a variety of ways to do that. One popular way is to ask the EIA, for example, to produce regular public information and updates as they have done in the past but specific to this policy which would give consumers the confidence to know that this is working to their advantage. Additionally as we just mentioned, the policy that gives the administration the ability to restrict exports if that makes sense in circumstances would do so.

Mr. CASTRO. Let me have 20 seconds.

Mr. BORDOFF. I just want to very quickly—I am sorry. I was just going to answer your question which is if the President allows exports under a national interest determination, he can revoke it. EPCA allows the immediate issuance of regulations to restrict exports without seeking public comments in the event of a true supply emergency and disruption. And if Congress repealed the ban it could give the President authority to re-ban, or the Secretary of Commerce could impose controls under the Export Administration Act, short supply controls, or potentially the President can invoke his emergency authority under the International Emergency Economic Powers Act to limit crude oil exports. So authorities exist in the event of a true emergency like the 1970s.

Mr. CASTRO. Okay, I am out of time but can I just make a final comment, Chairman? Thank you. That I think it is going to be difficult to lift that ban carte blanche. That there is going to have to be safeguards in place and I think we ought to consider whether there is an opportunity to also support alternative energies if this is going to happen. So thank you. I yield back.

Mr. POE. The gentleman yields back. The chair recognizes the gentleman from California, Mr. Rohrabacher, for his comments.

Mr. ROHRABACHER. Thank you very much, Mr. Chairman, and thank you, Mr. Chairman, for holding this hearing. Mr. Bordoff, in your testimony you mentioned several times about these trains, about dangerous long trains. No, but I think you did too and I caught that. Maybe I am mistaken.

Well, then I won't ask you the question. I will just suggest which is the next—well, let me ask you this then. Did your organization support the Keystone Pipeline? No, no, Mr. Bordoff.

Mr. BORDOFF. I am a professor at Columbia University. We don't take institutional positions on particular issues.

Mr. ROHRABACHER. Well, did you support the Keystone Pipeline?

Mr. BORDOFF. I have said that with Keystone, I think we should be focusing on the issues that really matter to achieve meaningful reductions in climate change and carbon emissions, and I think a decision in either direction on Keystone doesn't have a huge impact on affecting greenhouse gas emissions.

Mr. ROHRABACHER. Now, first of all, let me just note that—is it Kretzmann?

Mr. KRETZMANN. Yes, that is right. Kretzmann.

Mr. ROHRABACHER. Mr. Kretzmann, first of all, let me say that it is refreshing to have someone as open and honest as you are about your positions testifying before us. Most of the people who come here opposing the pipeline, the Keystone Pipeline, and supporting all of these various controls are not as open about what their goal really is, and you honestly have expressed you want us to end oil production. You are not in favor of any more oil production and would shut down oil production now if you could. And that is, frankly, refreshing to hear someone being this open about this because that is not what we get.

And in terms of your motive, I understand and appreciate it, the fact that you believe that production of CO2, which goes hand in glove with oil production in our country, that that is harmful to the world environment via a theory that CO2, increased CO2, will increase the temperature of the planet.

Let me just note, Mr. Chairman, I would ask unanimous consent at this point, to be put in the record, the name of 10 prominent scientists from around the United States who totally disagree with that particular theory that more CO2 means that there will be higher temperatures.

Mr. POE. Without objection it will be made part of the record.

Mr. ROHRABACHER. And so, and I know there are other scientists who disagree with that. So we have people who are prominent scientists on both sides of this issue, and I can't help but notice, however, that the predictions based on that particular theory haven't been coming true in the last 18 years based on—I mean I can remember the debate that we have had here, where several scientists were quoted as saying we are going to reach a tipping point, and the tipping point will be a major jump within a very short period of time of temperature. And not only have we not seen this tipping point and major jump in temperature, but there has been a basically no increase in temperature for 18 years, yet the CO2 rates have gone up.

So just my, the scientists that I am talking about as well as my common sense tells me, we shouldn't be basing policy or energy policy on that theory. And I respect the fact that you are an intelligent person and the people who you have spoken to are intelligent, but we have a difference of opinion on that. And I would think that trying to implement what you have honestly expressed, which I believe is not being honestly expressed by others, would mean a major decline in the standard of living of our people. And I would appreciate an honest discussion, so thank you for being here and being open in your testimony today. But I will give you 30 seconds to refute everything I said.

Mr. KRETZMANN. That may take a little bit longer but I will make some reference quickly. Thank you for your honesty, Congressman. On a variety of different things I would say that it is quite clear the majority of, the vast majority, 97 percent of climate scientists are completely clear on the dangers associated with climate change. And I would like to submit for the record the contents of skepticalscience.com, in which you will find the questions that are often posed by people who are questioning climate change and answers from climate scientists. So if that would be possible I would love to do that.

Re your question about not observing temperature changes, you should look at the temperature of water. Because the oceans have been absorbing the heat over the several decades and that is where the heat is going and they are pretty much done with absorbing the heat, now we are going to see the rest of it jump up very high. That is what the scientists tell me.

I do not want to bring oil production to zero immediately. That would be irresponsible and disastrous. However, it is clear that if we can, we will need to cut oil production and fossil fuel use down to essentially zero by 2050. That is a long time particularly for a

62

country as great as ours that can put a man on the moon and do anything we want to. I believe we can make this transition, and I believe we can do it in a way that will be healthy for our economy and great for our communities and we will all have a better standard of living at the end of it.

Mr. ROHRABACHER. Thank you for expressing that vision. I disagree with it but——

Mr. KRETZMANN. That is not a surprise.

Mr. ROHRABACHER [continuing]. I appreciate that. Thank you so much, Mr. Kretzmann.

Mr. KRETZMANN. Thank you, Mr. Chairman.

Mr. POE. I thank the gentleman. Yields back. The chair recognizes the ranking member for a final comment.

Mr. KEATING. Thank you, Mr. Chairman. I just, since my colleague and friend wanted to submit 10 different scientific references against climate change, I would like unanimous consent to put in a paper reflecting the 12,000 peer reviewed scientific articles, 97 percent of which indicate climate change exists as well.

Mr. POE. If you have those names they will be submitted to the record.

Mr. KRETZMANN. I can get names for you.

Mr. KEATING. And just in closing, thank you again, Mr. Chairman. Thank the witnesses. And I think each of you in your own way really gave important information and I appreciate your testimony. I would just say too there is another cost that, really, it is hard to quantify perhaps besides just the cost in the Northeast, the potential with the refineries contracting more, and that is the cost of climate change. I have the highest yielding dollar fish industry in that city of New Bedford that I represent and climate change has affected drastically the fishing industry.

And also in terms of the flooding and the erosion, our tourist industry is threatened and is threatened right now from that. And look at the cost of the historic snowfalls that the Northeast and my state in particular have suffered through. So I have left the cost of spills and mitigation of that and clean-ups as well. So there is costs all the way around and I think it is an important discussion to have. And I appreciate the witness and I appreciate the opportunity, Mr. Chairman, to have this hearing.

Mr. POE. I thank the gentleman. And just a final word. I am very concerned about the loss of jobs because of the recent developments, 50 percent of the rigs in the state of Texas have been shut down since Thanksgiving; 70,000 people have lost their jobs. I think it is important that we at least treat America the same way we treat the Iranians. If we lift the ban on exporting Iranian oil we ought to lift the ban on exporting American oil. I think it makes sense. It is a national security issue. It is also an energy issue, and it is a jobs issue as well.

But I do thank all of the panelists for being here and the members who have participated in this lively discussion. The subcommittee is adjourned.

[Whereupon, at 11:49 a.m., the subcommittee was adjourned.]

APPENDIX

MATERIAL SUBMITTED FOR THE RECORD

SUBCOMMITTEE HEARING NOTICE
COMMITTEE ON FOREIGN AFFAIRS
U.S. HOUSE OF REPRESENTATIVES
WASHINGTON, DC 20515-6128

Subcommittee on Terrorism, Nonproliferation, and Trade
Ted Poe (R-TX), Chairman

TO: MEMBERS OF THE COMMITTEE ON FOREIGN AFFAIRS

You are respectfully requested to attend an OPEN hearing of the Committee on Foreign Affairs, to be held by the Subcommittee on Terrorism, Nonproliferation, and Trade in Room 2172 of the Rayburn House Office Building (and available live on the Committee website at http://www.ForeignAffairs.house.gov):

DATE: Tuesday, April 14, 2015

TIME: 10:15 a.m.

SUBJECT: The Crude Oil Export Ban: Helpful or Hurtful?

WITNESSES: Panel I
                   The Honorable Joe Barton
                   United States House of Representatives

                   The Honorable Michael McCaul
                   United States House of Representatives

                   Panel II
                   Mr. Jason Grumet
                   Founder and President
                   Bipartisan Policy Center

                   Ms. Elizabeth Rosenberg
                   Director
                   Energy, Economics, and Security Program
                   Center for a New American Security

                   Mr. Jason Bordoff
                   Founding Director
                   Center on Global Energy Policy
                   Columbia University

                   Mr. Stephen Kretzmann
                   Founder and Executive Director
                   Oil Change International

By Direction of the Chairman

The Committee on Foreign Affairs seeks to make its facilities accessible to persons with disabilities. If you are in need of special accommodations, please call 202/225-5021 at least four business days in advance of the event, whenever practicable. Questions with regard to special accommodations in general (including availability of Committee materials in alternative formats and assistive listening devices) may be directed to the Committee.

COMMITTEE ON FOREIGN AFFAIRS

MINUTES OF SUBCOMMITTEE ON _____ *Terrorism Nonproliferation and Trade* _____ HEARING

Day ___*Tuesday*___ Date ___*April 14, 2015*___ Room ___*2172*___

Starting Time ___*10:15 a.m.*___ Ending Time ___*11:49 a.m.*___

Recesses [____] (____ to ____) (____ to ____) (____ to ____) (____ to ____) (____ to ____) (____ to ____)

Presiding Member(s)

Chairman Ted Poe

Check all of the following that apply:

Open Session ☑   Electronically Recorded (taped) ☑
Executive (closed) Session ☐   Stenographic Record ☑
Televised ☑

TITLE OF HEARING:

"The Crude Oil Export Ban: Helpful or Hurtful?"

SUBCOMMITTEE MEMBERS PRESENT:

Reps. Poe, Wilson, Issa, Cook, Perry, Ribble, Keating, Sherman, Castro, Kelly

NON-SUBCOMMITTEE MEMBERS PRESENT: *(Mark with an * if they are not members of full committee.)*

Rep. Rohrabacher

HEARING WITNESSES: Same as meeting notice attached? Yes ☑ No ☐
(If "no", please list below and include title, agency, department, or organization.)

STATEMENTS FOR THE RECORD: *(List any statements submitted for the record.)*

Reps. Perry and Rohrabacher (Statements for the Record)

TIME SCHEDULED TO RECONVENE _____
or
TIME ADJOURNED ___*11:49 a.m.*___

_____
Subcommittee Staff Director

Statement for the Record
Congressman Dana Rohrabacher
HFAC TNT Hearing: Crude Oil Exports
April 14, 2015

List of 100 scientists who agree that:
- The case for alarm regarding climate change is grossly overstated;
- Surface temperature changes over the past century have been episodic and modest;
- There has been no net global warming for over a decade;
- The computer models forecasting rapid temperature change abjectly fail to explain recent climate behavior; and
- Characterization of the scientific facts regarding climate change and the degree of certainty informing the scientific debate is simply incorrect.

SYUN AKUSOFU, PH.D
UNIVERSITY OF ALASKA

ARTHUR G.ANDERSON, PH.D
DIRECTOR OF RESEARCH, IBM (RETIRED)

CHARLES R.ANDERSON, PH.D
ANDERSON MATERIALS EVALUATION

J. SCOTT ARMSTRONG, PH.D
UNIVERSITY OF PENNSYLVANIA

ROBERT ASHWORTH
CLEARSTACK LLC

ISMAIL BAHT, PH.D
UNIVERSITY OF KASHMIR

COLIN BARTON
CSIRO (RETIRED)

DAVID J. BELLAMY, OBE
THE BRITISH NATURAL ASSOCIATION

JOHN BLAYLOCK
LOS ALAMOS NATIONAL LABORATORY (RETIRED)

EDWARD F. BLICK, PH.D
UNIVERSITY OF OKLAHOMA (EMERITUS)

SONJA BOEHMER-CHRISTIANSEN, PH.D
UNIVERSITY OF HULL

BOB BRECK
AMS BROADCASTER OF THE YEAR 2008

JOHN BRIGNELL
UNIVERSITY OF SOUTHAMPTON (EMERITUS)

MARK CAMPBELL, PH.D
U.S. NAVAL ACADEMY

ROBERT M. CARTER, PH.D
JAMES COOK UNIVERSITY

IAN CLARK, PH.D
PROFESSOR, EARTH SCIENCES, UNIVERSITY OF OTTAWA, OTTAWA, CANADA

ROGER COHEN, PH.D
FELLOW, AMERICAN PHYSICAL SOCIETY

PAUL COPPER, PH.D
LAURENTIAN UNIVERSITY (EMERITUS)

RICHARD S. COURTNEY, PH.D
REVIEWER, INTERGOVERNMENTAL PANEL ON CLIMATE CHANGE

UBERTO CRESCENTI, PH.D
PAST-PRESIDENT, ITALIAN GEOLOGICAL SOCIETY

SUSAN CROCKFORD, PH.D
UNIVERSITY OF VICTORIA

JOSEPH S. D'ALEO
FELLOW, AMERICAN METEOROLOGICAL SOCIETY

JAMES DEMEO PH.D
UNIVERSITY OF KANSAS (RETIRED)

DAVID DEMING, PH.D
UNIVERSITY OF OKLAHOMA

DIANE DOUGLAS, PH.D
PALEOCLIMATOLOGIST

DAVID DOUGLASS, PH.D
UNIVERSITY OF ROCHESTER

CHRISTOPHER ESSEX, PH.D
UNIVERSITY OF WESTERN ONTARIO

JOHN FERGUSON, PH.D
UNIVERSITY OF NEWCASTLE UPON TYNE (RETIRED)

MICHAEL FOX, PH.D
AMERICAN NUCLEAR SOCIETY

GORDON FULKS, PH.D

GORDON FULKS AND ASSOCIATES

LEE GERHARD, PH.D
STATE GEOLOGIST, KANSAS (RETIRED)

GERHARD GERLICH, PH.D
TECHNISCHE UNIVERSITAT BRAUNSCHWEIG

IVAR GIAEVER, PH.D
NOBEL LAUREATE, PHYSICS

ALBRECHT GLATZLE, PH.D
SCIENTIFIC DIRECTOR, INTTAS (PARAGUAY)

WAYNE GOODFELLOW, PH.D
UNIVERSITY OF OTTAWA

JAMES GOODRIDGE
CALIFORNIA STATE CLIMATOLOGIST (RETIRED)

LAURENCE GOULD, PH.D
UNIVERSITY OF HARTFORD

VINCENT GRAY, PH.D
NEW ZEALAND CLIMATE COALITION

WILLIAM M. GRAY, PH.D
COLORADO STATE UNIVERSITY

KENNETH E. GREEN, D.ENV.
AMERICAN ENTERPRISE INSTITUTE

KESTEN GREEN, PH.D
MONASH UNIVERSITY

WILL HAPPER, PH.D
PRINCETON UNIVERSITY

HOWARD C. HAYDEN, PH.D
UNIVERSITY OF CONNECTICUT (EMERITUS)

BEN HERMAN, PH.D
UNIVERSITY OF ARIZONA (EMERITUS)

MARTIN HERTZBERG, PH.D.
U.S. NAVY (RETIRED)

DOUG HOFFMAN, PH.D
AUTHOR, THE RESILIENT EARTH

BERND HUETTNER, PH.D

OLE HUMLUM, PH.D
UNIVERSITY OF OSLO

NEIL HUTTON
PAST PRESIDENT, CANADIAN SOCIETY OF
PETROLEUM GEOLOGISTS

CRAIG D. IDSO, PH.D
CENTER FOR THE STUDY OF CARBON DIOXIDE AND
GLOBAL CHANGE

SHERWOOD B. IDSO, PH.D
U.S. DEPARTMENT OF AGRICULTURE (RETIRED)

KIMINORI ITOH, PH.D
YOKOHAMA NATIONAL UNIVERSITY

STEVE JAPAR, PH.D
REVIEWER, INTERGOVERNMENTAL PANEL ON
CLIMATE CHANGE

STEN KAIJSER, PH.D
UPPSALA UNIVERSITY (EMERITUS)

WIBJORN KARLEN, PH.D
UNIVERSITY OF STOCKHOLM (EMERITUS)

JOEL KAUFFMAN, PH.D
UNIVERSITY OF THE SCIENCES,
PHILADELPHIA (EMERITUS)

DAVID KEAR, PH.D
FORMER DIRECTOR-GENERAL, NZ DEPT.
SCIENTIFIC AND INDUSTRIAL RESEARCH

RICHARD KEEN, PH.D
UNIVERSITY OF COLORADO

DR. KELVIN KEMM, PH.D
LIFETIME ACHIEVERS AWARD, NATIONAL SCIENCE
AND TECHNOLOGY FORUM, SOUTH AFRICA

MADHAV KHANDEKAR, PH.D
FORMER EDITOR, CLIMATE RESEARCH

ROBERT S. KNOX, PH.D
UNIVERSITY OF ROCHESTER (EMERITUS)

JAMES P. KOERMER, PH.D
PLYMOUTH STATE UNIVERSITY

GERHARD KRAMM, PH.D
UNIVERSITY OF ALASKA FAIRBANKS

WAYNE KRAUS, PH.D
KRAUS CONSULTING

OLAV M. KVALHEIM, PH.D
UNIV. OF BERGEN

ROAR LARSON, PH.D
NORWEGIAN UNIVERSITY OF SCIENCE
AND TECHNOLOGY

JAMES F. LEA, PH.D

DOUGLAS LEAHY, PH.D
METEOROLOGIST

PETER R. LEAVITT
CERTIFIED CONSULTING METEOROLOGIST

DAVID R. LEGATES, PH.D
UNIVERSITY OF DELAWARE

RICHARD S. LINDZEN, PH.D
MASSACHUSETTS INSTITUTE OF TECHNOLOGY

HARRY F. LINS, PH.D.
CO-CHAIR, IPCC HYDROLOGY AND
WATER RESOURCES WORKING GROUP

ANTHONY R. LUPO, PH.D
UNIVERSITY OF MISSOURI

HOWARD MACCABEE, PH.D, MD
CLINICAL FACULTY, STANFORD MEDICAL SCHOOL

HORST MALBERG, PH.D
FREE UNIVERSITY OF BERLIN

BJORN MALMGREN, PH.D
GOTEBURG UNIVERSITY (EMERITUS)

JENNIFER MAROHASY, PH.D
AUSTRALIAN ENVIRONMENT FOUNDATION

ROSS MCKITRICK, PH.D
UNIVERSITY OF GUELPH

PATRICK J.MICHAELS, PH.D
UNIVERSITY OFVIRGINIA

TIMMOTHY R.MINNICH,MS
MINNICH AND SCOTTO, INC.

ASMUNN MOENE, PH.D
FORMER HEAD, FORECASTING
CENTER,METEOROLOGICAL INSTITUTE, NORWAY

MICHAEL MONCE, PH.D
CONNECTICUT COLLEGE

DICK MORGAN, PH.D
EXETER UNIVERSITY (EMERITUS)

NILS-AXEL MÖRNER, PH.D
STOCKHOLM UNIVERSITY (EMERITUS)

DAVID NOWELL, D.I.C.
FORMER CHAIRMAN, NATO
METEOROLOGY CANADA

CLIFF OLLIER, D.SC.
UNIVERSITY OFWESTERN AUSTRALIA

GARTH W. PALTRIDGE, PH.D
UNIVERSITY OF TASMANIA

ALFRED PECKAREK, PH.D
ST. CLOUD STATE UNIVERSITY

DR. ROBERT A. PERKINS, P.E.
UNIVERSITY OF ALASKA

IAN PILMER PH.D
UNIVERSITY OF MELBOURNE (EMERITUS)

BRIAN R. PRATT, PH.D
UNIVERSITY OF SASKATCHEWAN

JOHN REINHARD, PH.D
ORE PHARMACEUTICALS

PETER RIDD, PH.D
JAMES COOK UNIVERSITY

CURT ROSE, PH.D
BISHOP'S UNIVERSITY (EMERITUS)

PETER SALONIUS M.SC.
CANADIAN FOREST SERVICE

GARY SHARP, PH.D

CENTER FOR CLIMATE/OCEAN
RESOURCES STUDY

THOMAS P. SHEAHAN, PH.D
WESTERN TECHNOLOGIES, INC.

ALAN SIMMONS
AUTHOR, THE RESILIENT EARTH

ROY N. SPENCER, PH.D
UNIVERSITY OF ALABAMA—HUNTSVILLE

ARLIN SUPER, PH.D
RETIRED RESEARCH METEOROLOGIST, U.S. DEPT.
OF RECLAMATION

EDUARDO P. TONNI, PH.D
MUSEO DE LA PLATA (ARGENTINA)

RALF D.TSCHEUSCHNER, PH.D

DR.ANTON URIARTE, PH.D
UNIVERSIDAD DEL PAISVASCO

BRIAN VALENTINE, PH.D
U.S. DEPARTMENT OF ENERGY

GOSTA WALIN, PH.D
UNIVERSITY OF GOTHENBURG (EMERITUS)

GERD-RAINERWEBER, PH.D
REVIEWER, INTERGOVERNMENAL PANEL ON
CLIMATE CHANGE

FORESE-CARLOWEZEL, PH.D
URBINO UNIVERSITY

EDWARD T.WIMBERLEY, PH.D
FLORIDA GULF COAST UNIVERSITY

MIKLOS ZAGONI, PH.D
REVIEWER, INTERGOVERNMENTAL PANEL ON
CLIMATE CHANGE

ANTONIO ZICHICHI, PH.D
PRESIDENT, WORLD FEDERATION OF SCIENTISTS

Statement for the Record
Congressman Scott Perry
HFAC TNT Hearing: Crude Oil Exports
April 14, 2015

The original economic rationale for crude export restrictions no longer applies. Today's oil market looks very different than in the 1970s when current crude oil export restrictions were first put in place. At that time, the US had adopted domestic price controls to combat inflation and crude export restrictions were necessary to make those price controls effective. While price controls have long since fallen away, crude export restrictions remain.

Increased US crude oil exports also has a significant geopolitical component. Currently, the international oil market is vulnerable to the influence of Russia, Iran, Venezuela, and others. Allowing American exports would reduce those countries' ability to use oil as a foreign policy weapon. For example, Japan and South Korea rely on crude oil from Iran to satisfy their growing energy consumption. The U.S. can help diminish that reliance. Lifting the export ban would offer greater energy sources for our allies, both in Asia and beyond.

In summary, I see no sensible reason—neither economically nor geopolitically—to maintain the ban on crude oil exports